A Handbook of Agriculture in India

A Handbook of Agriculture in India

Mahipal Singh Rana

CYBER TECH PUBLICATIONS
4264/3, Ansari Road, Daryaganj, New Delhi-110002 (India)
Ph.: 011-23244078. Fax: 011-23280028
E-Mail: cyberpublicationsdelhi@yahoo.com

A HANDBOOK OF AGRICULTURE IN INDIA

Mahipal Singh Rana

First Edition 2009

Published by :

G.S. Rawat for **Cyber Tech Publications**
4264/3, Ansari Road, Daryaganj, New Delhi-110002 (India)
Ph.: 011-23244078. Fax: 011-23280028
E-Mail: cyberpublicationsdelhi@yahoo.com

Printed at : **Amit Enterprises**, Maujpur, Delhi, (9810878184)

Preface

India's record of progress in agriculture over the past four decades has been quite impressive. The agriculture sector has been successful in keeping pace with rising demand for food. The contribution of increased land area under agricultural production has declined over time and increases in production in the past two decades have been almost entirely due to increased productivity. Contribution of agricultural growth to overall progress has been widespread. Increased productivity has helped to feed the poor, enhanced farm income and provided opportunities for both direct and indirect employment. The success of India's agriculture is attributed to a series of steps that led to availability of farm technologies which brought about dramatic increases in productivity in 70s and 80s often described as the Green Revolution era. The major sources of agricultural growth during this period were the spread of modern crop varieties, intensification of input use and investments leading to expansion in the irrigated area. In areas where 'Green Revolution' technologies had major impact, growth has now slowed.

This book is about the agriculture and its patterns in India, I hope that readers will find it of great significance.

—Author

CONTENTS

1

Introduction

India has made lot of progress in agriculture since independence in terms of growth in output, yields and area under many crops. It has gone through a green revolution, a white revolution, a yellow revolution and a blue revolution. Today, India is the largest producer of milk, fruits, cashewnuts, coconuts and tea in the world, the second largest producer of wheat, vegetables, sugar and fish and the third largest producer of tobacco and rice. The per capita availability of foodgrains has risen in the country from 350 gm in 1951 to about 500 gm per day now, of milk from less than 125 gm to 210 gm per day and of eggs from 5 to 30 per annum despite the increase in population from 35 crores to 95 crores. However, India would have been in an even better position now both in terms of agricultural output and economic development had our planners given the required importance to its development it deserved in the early years since independence.

Even today, the farmers in India are able to obtain only 15 per cent of their requirements of agricultural credit from banks. The various state seed corporations are able to produce only 1 0 per cent of the seeds required by our farmers. At present only 2330 per cent of the farmers are able to derive any benefits of extension services provided by various government agencies and every year about 20 per cent of the crop is lost due to mishandling, spillage, floods, droughts and pests and diseases.

In fruits and vegetables the loss is around 30 per cent. The farmers are still being exploited by moneylenders who provide finance at exorbitant interest rates and there are cartels of traders who pay very little for their produce even in the well recognized mandies in the country.

GROSS CAPITAL FORMATION IN AGRICULTURE

Though the overall growth of Indian economy has depended much upon the performance of agriculture, over the years, not much public investment has been made on its development. There is a steady deceleration in public, investment in gross capital formation in agriculture. In 1980-81, the public investment as a percentage of gross capital formation in agriculture was 38.7 percent which fell to 16.2% in 1996-97. During this period the share of private investment, however, rose from 61.3% to 83.8%.

This rise is attributed mostly to better terms of trade offered by the government to agriculture vis-a-vis industry. The fall in public sector investment is attributed to increase in current expenditure to meet higher subsidies on food, fertilizers, electricity, irrigation, credit and other r farm inputs rather than creating assets. In recent years public expenditure on maintenance of existing irrigation projects has gone up and investment on new projects has fallen. Many of the projects started in early 1980's are still awaiting completion because of cost overruns and non-availabilty of necessary timely funds.

In the 8th plan it was proposed to raise the investment in agriculture 10 18.7% if the total plan outlay, but the actual investment turned out to be about 10-11% only.

GROWTH IN FOODGRAIN PRODUCTION

The foodgrain production in the 1950s rose in India because of expansion in area. In the 1960s, the growth rate was very poor which resulted in large scale imports. The development, production and use of better seeds increased the productivity of wheat in 1970s and rice in 1980s. The eighties was also a period of green revolution which enabled India to become self-sufficient in foodgrain production and even a marginal exporter. In the Nineties, however, the annual growth rate in production has fallen to 1.66 per cent from 3.54 per cent recorded in the eighties.This is a matter of serious concern for the country as this growth rate is just matching the annual growth rate of population and 40 percent of the population in the country is still living below the poverty line.

The production of foodgrains in the country fell to 192.4 million tonnes in 1997-98 from 199.4 million tonnes in 1996-97.The production of wheat dropped by 3.5 million tonnes and coarse cereals by 3 million tonnes. The kharif 1998-99 crop was also no better. However, the 1998-99 Rabi crop is expected to be good with a 5.2 million tonnes higher foodgrain output expected from it. The crop is estimated at 96.5 million tonnes.

Gross Capital Formation in Agriculture (At 1993-94 prices in Rs. Crores)

Gross Capital Formation in Agriculture (At 1993-94 prices in Rs. Crores)					
Year	Gross Capital Formation			Per Cent Share	
	Public	Private	Total	Public	Private
1993-94	4468	11377	15845	28.2	71.8
1994-95	4970	13244	18214	27.3	72.7
1995-96	4776	15168	19944	23.9	76.1
1996-97	4347	15555	19902	21.8	78.2
1997-98	4416	16579	20995	21.0	79.0

Annual Growth in Foodgrain Production (% per annum compounded)					
Crop	1950-51 to 1959-60	1960-61 to 1969-70	1970-71 to 1979-80	1980-81 to 1989-90	1990-91 to 1997-98
Rice	3.28	8.05	1.91	4.29	1.53
Wheat	4.51	5.90	4.69	4.24	3.67
Coarse Grains	2.75	1.48	0.74	0.74	(-)0.49
All Cereals	3.0	2.51	2.37	3.63	1.84
Pulses	2.72	1.35	(-)0.54	2.78	0.76
All Foodgrains	3.22	1.72	2.08	3.54	1.66

Agricultural ministry has projected a 1998-99 ouputof foodgrains at more than 200 million tonnes. The union government had plans to cover about 2 million farmers in the country with a total credit of Rs. 168 crores in 1999 with'kisan cards'. It has fixed interest rates on creditgiven to farmers through these cards at 13.26 per cent including tax on interest on the first Rs. 2 lakhs credit and , at 6.02 percent on amounts above Rs. 2.00 lakhs.

NATIONAL SAMPLE SURVEY

According to a National Sample Survey (NSS), the monthly per capita consumption of cereals in India has declined from 14.4 kg to 12.8 kg in rural areas over the period 1987-1997. This is despite the fact that at constant prices the monthly per capita consumer expenditure (MPCE) has risen by 6 per cent in rural areas and by 13 per cent in urban areas over the same period.

The all-India MPCE during 1997 was Rs. 459, out of which expenditure on food was Rs. 255 and on non-food items Rs. 204. In rural areas, the MPCE was Rs. 395, with expenditure on food items Rs. 232 and non-food items Rs. 163, while in urban areas the MPCE was Rs. 645, expenditure on food items Rs. 320 and non-food items Rs. 325. The urban-rural disparity in MPCE was only 12 per cent in Punjab but 100 per cent in Maharashtra. The average household size in the country was 5.0 in rural areas and 4.6 in urban areas in 1997.

NNMB STUDY

According to a survey conducted by the National Nutrition Monitoring Bureau (NNMB) in 10 states (A.P., Gujarat, Karnataka, Kerala, M.P., Maharashtra, Orissa, T.N., W.B. and U.P.), the average consumption of cereals and millets was slightly more than or equal to the recommended dietary allowance (RDA) level, though it was showing a failing tendency with years. The intake of pulses and legumes has remained lower than RDA level. The average intake of leafy vegetables, an inexpensive source of vitamins and minerals, was only one-third of RDA level, while the consumption of other vegetables was marginally lower than RDA levels. The consumption of fruits has risen slightly during 1990-1997. The daily intake of milk and milk products, visible fat, and sugar and jaggery was only half, half, and two-thirds of RDA levels respectively in 1997. The intake of proteins was comparable to RDA but has fallen marginally since,1990. In terms of calorific value of food, the intake is failing since 1982 and is below RDA. The intake of iron is apparently normal but the bio-availability of iron in Indian diet is poor at 3 per cent. The intake of vitamin A varies from 30 to 60 per cent of RDA in diets in different regions. The intake of riboflavin is about two-thirds of RDA level. The intake of vitamin C is satisfactory.

SEARCH FOR AGRICULTURAL POLICY AND PROGRAMMES

The Institute of Development Studies, Jaipur has taken up a comprehensive all-India research project on agricultural reforms 'Equity-driven trade and marketing policy strategies for improved performance of Indian Agriculture' with financial support from the Australian government. The project will work out a package of agricultural policies and programmes to improve the efficient utilization of resources, food security for masses and reduction in regional disparities.

At present the non-availability of more land is a major constraint in the growth of agriculture. With no scope for expansion in the net sown area, there is a need to increase the cropping intensity from 134.2 to 140.4 per cent, the value of output from Rs. 15,326 to Rs. 17,688 per hectare (based on 1996-97 prices), the gross cropped area from 190.5 to 199.3 million hectares and fertilizer consumption from 14.34 to 20.00 million tonnes.

There is a need to give more attention to high growth areas like animal husbandry, dairy and fishery and also improve, expand and upgrade the marketing, storage and distribution infrastructure and set up facilities for packaging, grading and certification of agricultural commodities.

CENTRAL PLAN OUTLAY

In the 1998-99 budget, the allocation for agriculture was hiked by 58 per cent to Rs. 2,854 crores from Rs. 1,807 crores in 1997-98 by enhancing outlays on irrigation and agricultural credit. The outlay on watershed development programmes was raised from Rs. 517 crores in 1997-98 to Rs. 677 crores in 1998-99. All the watershed development programmes implemented by different ministries were proposed to be unified to make a better utilization of funds. The allocation for accelerated irrigation programme was enhanced by Rs. 200 crores.

In 1998-99, the share capital of NABARD was raised from Rs. 1,500 crores to 2,000 crores with Rs. 100 crores contributed by the exchequerand Rs.400 crore by the RBI. This was to enable NABARD increase its leverage and obtain additional resources from the market to meet its credit needs for agriculture. A sum of Rs. 264 crores was also provided for rehabilitation and recapitalisation of regional rural banks. A provision of Rs. 60 crore was made for technology mission on cotton, Rs. 100 crore for experimental crop insurance scheme and Rs. 500 crores for rural infrastructure development fund of NABARD. In the budget, the subsidy on potassium and phosphatic fertilizers was increased, the import duty on biopesticides was reduced, excise duty on jute products was cut, 100 per cent tax-exemption on bio-pesticide manufacturing units was granted, duty on diesel generating sets of upto 1 0 HP was reduced and the manufacture of agricultural implements in the small scale sector was dereserved to improve their quality.

Gross Capital Formation in Agriculture (At 1980-81 prices in Rs. Crores)

Year	Gross Capital Formation			Per Cent Share	
	Public	Private	Total	Public	Private
1960-61	589	1079	1668	35.3	64.7
1970-71	789	1969	2758	28.6	71.4
1980-81	1796	2840	4636	38.7	61.3
1990-91	1154	3440	4594	25.1	74.9
1991-92	1002	3727	4729	21.2	78.8
1992-93	1061	4311	5372	19.7	80.3
1993-94	1153	3885	5038	22.9	77.1
1994-95	1316	4949	6256	21.0	79.0
1995-96	1268	5693	6961	18.2	81.8
1996-97	1132	5867	6999	16.2	83.8

Which fell to 16.2 percent in 1996-97. During this period the share of private investment, however, rose from 61.3 per cent to 83.8 per cent.This rise is attributed mostly to better terms of trade offered by the government to agriculture vis-a-vis industry.The fall in public sector investment is attributed to increase in current expenditure to meet higher subsidies on food, fertilisers, electricity, irrigation, credit and other farm inputs rather than creating assets. In recent years public expenditure on maintenance of existing irrigation projects has gone up and investment on new projects has fallen. Many of the projects started in early 1980's are still awaiting, completion because of cost overruns and non-availability of necessary timely funds. In the Eighth Plan it was proposed to raise the investment in agriculture to 18.7 percentof the total plan outlay but the actual investment turned out to be about 10 to 11 per cent only.

REGULATED MARKET MARKET AGRICULTURAL PRODUCE				
Year	Regulated Markets (Nos.)	Commodities Under Grading Standards (Nos.)	Cold Storages	
			(Nos.)	Capacity (Mil Tonnes)
1990-91	6640	142	2,930	7.68
1991-92	6738	143	2,973	7.78
1992-93	6772	148	3,053	8.09
1993-94	6809	150	3,124	8.17
1994-95	6836	151	3,167	8.56
1995-96	6968	1533	3,253	8.73
1996-97	7062	162	-	-

The government also announced a scheme to issue 'Kisan Credit Cards'to farmers.Under the scheme the farmers will be able to obtain loans from designated banks on the basis of their land holdings for the purchase of seeds, Fertilizers,pesticides and any other inputs as and when they are needed.

The scheme will be open to about 1 00 lakh farmer farlifies. The procedural details of the scheme were to be worked out and announced by NABARD after working out the modalities. The necessary refinance to the commercial and other banks for the implementation of the scheme was to be provided by NABARD.

Of the budgetary allocation of Rs. 2,854 crores in 1998-99, however, the ministry of agriculture is expected to have spent only Rs. 1,974 crores in 1998-99 (revised estimates). For 1999-2000, the central plan outlay for agriculture and rural development was raised by 47 per cent over the revised estimate of 1998-99 to Rs. 2,897 crores. The lesser expenditure in 1998-99 is attributed to long delays in execution of work plans by states.

NINTH PLAN PROJECTIONS

According to an earlier estimate, by the end of the Ninth Plan, the demand for foodgrains in India is expected to rise to 220 million tonnes, 185 million tonnes for household consumption, 27.5 million tonnes for food processing industries and 7.5 million tonnes for export. To meet this demand, the area finder foodgrains was to be raised to 120 million hectares besides affecting the usual increases in the yield through the use of better quality seeds and effective pest and disease control.The demand for fruits was expected to go up to 50 million tonnes and, vegetables 95 million tonnes, edible oils 10 million tonnes and sugar 17 million tonnes. The foodgrain and edible oil target,however, is not expected to be achieved.

According to the Planning Commission, there was a significant shortfall in public sector. Investment in agriculture in the Eighth Plan which should be reversed in the Ninth Plan. The target for agricultural growth should be fixed taking into the account the gap in the domestic availability and the demand (both for domestic consumption and exports).

In value terms the output of agricultural and allied products should grow at 3.52 per cent to meet the domestic needs and to take advantage of the world market which offers significant opportunity to earn foreign exchange through exports, the annual growth rate should be 4.5 per cent. This would call for an investment of Rs. 2,68,300 crores at 1996-97 prices in the Ninth Plan which is 75 per cent higher than in the Eighth Plan.

AGRICULTURAL MARKETING

At present, most of the agricultural produce in the country is freely marketed through private trade operating in organized markets/mandies. Except for commodities whose prices are administered, viz., petroleum, coal, nitrogenous fertilizers etc, most agricultural commodity markets operate under the normal forces of demand and supply. In order to save the farmers from cartels of traders and encourage farmers to undertake the cultivation of certain specific crops (foodgrains, oilseeds, cotton, jute, tobacco, sugarcane, etc) the government also fixes minimum support/statutory prices for some crops and makes arrangements for their purchase on state account whenever their price fall below the support level.

The role of the government normally is limited to protecting the interests of producers and consumers, only in respect of wage goods, mass consumption goods and essential goods. The government is promoting organized marketing of agricultural commodities in the country. To achieve this, the governments of many states and union territories have enacted necessary legislation for regulation of agricultural produce markets. The number of regulated markets in the country went up from 6,640 in 1990-91 to 6,968 in 1995-96 and 7,062 in 1996-97.

The union government has also set up a number of organizations, under the aegis of ministry of agriculture and ministry of commerce, such as commodity boards, cooperative federations and export promotion councils for monitoring and boosting the production, consumption, marketing and export of various agricultural commodities.

These include the Commission for Agricultural Costs and. Prices (CACP) for recommending minimum prices of certain commodities; the Food Corporation of India Ltd. (FCI), the Cotton Corporation of India Ltd. (CCI), the Jute Corporation of India Ltd. (JCI), the National Cooperative Development Corporation Ltd. (NCDC), the National Cooperative Marketing Federation Ltd. (NAFED), the National Tobacco Growers Federation Ltd. (NTGF), the Tribal Cooperative Marketing Development Federation Ltd. (TRIFED), the National Consumers Cooperative Federation Ltd. (NCCF), etc for procurement and distribution of commodities; and the Tea Board, Coffee Board, Coir Board, Rubber Board, Tobacco Board, Spices Board, Coconut Board, Central Silk Board, the National Dairy Development Board (NDDB), State Trading Corporation

(STC), Agricultural & Processed Foods Export Development Authority (APEDA), Marine Products Export Development Authority (MPEDA), the Indian Silk Export Promotion Council, the Cashewnuts Export Promotion Council of India (CEPC), etc for promotion of production and exports of specific commodities.

The National Cooperative Union of India is an apex institution for cooperative Movement covering all types of cooperatives, It promotes and develops cooperatives by assisting groups of people in forming cooperatives. It also imparts cooperative education and organizes training programmes for managers and executives of cooperatives.

SHARE OF EXPORT OF AGRICULTURALPRODUCTS IN TOTAL EXPORTS(Rs.Crores)			
Year	**Total Exports**	**Agricultural Exports**	**% share**
1992-93	53,668	7,884	14.7
1993-94	69,751	13,021	18.7
1994-95	82,674	13,710	16.6
1995-96	1,06,353	21,136	19.8
1996-97	1,18,817	24,241	20.4
1997-98	1,26,290	23,690	18.8

NAFED is an apex cooperative marketing organization involved in procurement, distribution, export and import of select agricultural commodities. It is a central nodal agency of the government for undertaking price support operations for nonperishable commodities like pulses and oilseeds and for market intervention operations for perishable commodities like potatoes, onions, grapes, kinno, oranges, apples, eggs, chillies, black pepper, etc. During Jan'97-Feb'98, NAFED undertook market intervention in 13 commodities.

Most of the states in the country have set up Agricultural Produce Marketing Boards and Civil Supplies Corporations to deal with marketing and distribution of various commodities which are produced or sold in specific regions in the country. The union government provides financial and technical assistance to states in creation of infrastructural facilities in the markets including setting up of godowns in rural areas. Till the end of March 1997, grading standards have been prepared for 162 agricultural and allied commodities under the Agriculture Produce (Grading and Marketing) Act, 1937.

In the Ninth Plan period, the Tamil Nadu government is expected to spend Rs. 100 crores on extension of agricultural marketing facilities in the state.

The Maharashtra State Agricultural Marketing Board has plans to invest Rs. 514 crores in setting up of market yards and godowns for agricultural and floricultural produce in the Ninth Plan period. The state government has prepared a master plan for the integrated development of 1,930 markets including 405 rural markets in the state. For funding the development of 723 regulated markets, it is proposed to obtain 50 per cent of the funds from the union government at 12 per cent interest repayable in 14 years after a moratorium of two years, 25 per cent will be provided by the state government and 25 per cent by the Agriculture Produce Marketing Committee.

INDIA'S EXPORT OF SOME AGRO PRODUCTS DURING APRIL-SEPTEMBER, 1998 ($Million)

Description	Exports in April-September		
	1997	1998	% Growth
Tea	189.41	253.28	33.72
Coffee	252.62	267.30	5.81
Rice	399.38	627.14	57.03
Wheat	0.10	0.37	271.90
Other Cereals	2.12	0.68	-67.96
Pulses	31.36	30.05	-5.00
Raw Tobacco	106.09	88.10	-16.96
Spices	203.36	192.60	-5.29
Cashewnuts & CNSL	210.07	189.30	-9.89
Sesame, Nigerseed	38.95	24.24	-37.76
Groundnut	68.14	11.37	-83.31
Oil Meals	252.54	157.85	-37.50
Guar gum and Meal	67.14	100.07	49.05
Castor Oil	86.63	101.46	17.11

Shellac	6.29	4.48	-28.82
Sugar and Mollasses	58.49	1.06	-98.19
Processed Foods, Vegetables	75.58	67.01	11.33
Fruits and Vegetable Seeds	6.85	9.18	19.48
Floriculture	12.27	11.73	-4.36
Other processed Foods	109.30	88.10	-19.40
Meat and Meat Products	15.09	9.04	-40.06
Poultry and Dairy Products	11.99	7.33	-38.86
Spirits and Beverages	539.49	511.99	-5.10
Marine Products	180.88	31.54	-82.56
Raw Cotton	73.18	83.59	14.22
Natural Silk Textiles	51.70	41.91	-18.94
Wool and Woolen Products	35.29	36.67	3.90
Coir and Coir Products	77.66	51.58	-33.59
Jute Products	9.82	7.48	-23.81

The Kerala government has undertaken a Rs. 72 crore EU-assisted project to set up 6 wholesale markets specifically for marketing of Fruits and Vegetables agricultural produce. Of these 3 will be in urban areas and 3 in In India more than 100 million tonnes of fruits and vegetable rural areas.

The urban areas identified include Anayara in are produced and marketed through a number of rural, assisted project to set up 6 wholesale markets specifically for marketing of Fruits and Vegetables agricultural produce. Of these 3 will be in urban areas and 3 in In India more than 100 million tonnes of fruits and vegetable rural areas. The urban areas identified include Anayara in Thiruvananthapuram, Muvattapuzha in Ernakulam and S Batheri in Wayanad. In recent years because of new p options coming up in various states with high yielding pi materials, the monopoly of Kerala is being lost in the case number of crops like coconut, rubber, pepper, ginger, mom, cashewnut, etc.

At present about 60 per cent of the net sown area in Kerala is under rubber and coconut. The fall in their prices in 1998 affected the lives of nearly two-third of its agricultural workforce.

MARKETING OF FRUIT AND VEGETABLES

In India more than 100 million tonnes of fruits and vegetables are produced and marketed through a number of rural,wholesale and terminal markets every year. The market reforms in agricultural commodities have so far been limited mainly to foodgrains only. The marketing of fruits and vegetables has so far received little attention of the government. At present, there are a large number of intermediaries in this trade between the producer and consumer which has resulted in a wide gap in the producer and consumer price of these commodities which needs to be reduced to enable farmers receive remunerative prices for their produce and boost their production and consumption in the country.

INDIA'S EXPORTS OF FOOD PRODUCTS						
Description	**1991-92**	**1992-93**	**1993-94**	**1994-95**	**1995-96**	**1996-97**
Processed foods and vegetables	194	263	318	409	479	460
Animal Products	248	314	345	448	686	814
Other Processed Foods	248	285	512	411	1043	1642
Cereal Products	755	898	1280	1280	4952	3879
Marine Products	1376	1767	2504	3576	3501	4121
Total	2821	3527	4960	6124	10661	10916

The union government has not made any common regulation for the marketing of fruits and vegetables applicable all over the country, however, some of the state governments have enacted laws and Acts. The marketing of fruits and vegetables is under regulation in Rajasthan, Maharashtra, Bihar, Delhi, Uttar Pradesh and Karnataka and outside any regulatory purview in Tamil Nadu, West Bengal and Jammu & Kashmir.

At present, the wholesale markets for fruits and vegetables o a country-wide basis are concentrated in 10 large cities viz Delhi, Calcutta, Bangalore, Chennai, Mumbai, Jaipur, Nagpur Vijayawada, Lucknow and Varanasi. These cities account for the arrival of 75 per cent of vegetables marketed in major urban areas in India. Delhi, Calcutta, Mumbai and Pune alone account for the arrival of 55 per cent of vegetables. Delhi and Calcutta account for transit trade in 40 per cent fruits and 13 per cent vegetables. Delhi, Calcutta and Mumbai receive 59 per cent of total fruits and 46 per cent of vegetables.

In a study carried out in 1998, it was found that cities with more than 20 lakh population account for transit trade in 66 per cent of fruits and vegetables. In 48 cities with a population of more than 5 lakhs (as per 1981 census), there were 102 fruit and vegetable markets of which 54 were regulated. The wholesale trade in fruits takes place in 65 markets and in vegetables in 81 markets. Each market on an average serves a population of about 7 lakhs. The prices in these markets, however, are governed by demand and supply principles with the regulating market committee not playing any role in price correction or arresting wide price fluctuations.

FUTURES TRADING

In continuation of the process of economic reforms, resumption of 'Futures Trading' is receiving a renewed focus in the country. The Kabra Committee, in its 1994 report, had recommended the allowing of futures trading in 17 commodities including kapas; raw jute and jute goods; seed, oil and oilcakes of groundnut, rapeseed/mustard, cottonseed, sesame, sunflower, copra and soyabean; ricebran oil; linseed; onions; and silver.

INDIA'S EXPORT OF SOME AGRO PRODUCTS DURING APRIL-DECEMBER, 1998($Million)

Description	Exports in Apr.-Sept.		
	1997	1998	% Growth
Tea	313.24	390.05	24.52
Coffee	319.24	307.37	-3.72
Rice	580.51	1,092.25	88.15
Wheat	0.11	0.64	482.32
Spices	2.12	0.68	-67.96
Tobacco	106.09	88.10	-16.96
Sugar and Mollasses	68.49	3.54	-94.83
Fresh fruits and vegetables	108.30	91.53	-15.48
Marine Products	935.05	802.98	-14.12
Cotton (raw and waste)	31.63	30.05	-5.00

Futures trading in gur, potato, raw jute, turmeric, pepper and castorseed has been going on since long in the country. More recently the union government has allowed futures trading in coffee, castor oil, jute sacking and cotton and has also permitted futures trading in palmolein, oilseeds and oilmeals subject to completion of certain formalities. The futures trading in cotton was resumed after a gap of about 32 years at the Cotton Exchange Building in Mumbai on December 6, 1998. An international futures exchange in pepper has started operating in the country and for castor was expected to begin in May 1999 in Navi Mumbai.

COLD STORAGES

The number of cold storages (of more than 10 tonnes in the country has gone up from 2,930 at the end of 1990-91 to 3,253 at the end of 1995-96 and their capacities from 7.68 million tonnes to 8.73 million tonnes. The cold storages are being mostly used for the storage of agricultural commodities with potato being the largest stored commodity. At present setting up of cold storages in India is not a lucrative activity because of high investment and low rental charges prevailing in the country. In some of the states there are curbs on rental charges as they are fixed by the governments. The NCDC provides financial assistance for setting up cold storages in the cooperative sector. Till the end of March 1996, the NCDC had provided a total assistance of Rs. 79.97 crores for setting up of 248 cold storages with an installed capacity of 7.39 lakh tonnes. In 1997, the union government repealed the Cold Storage Order, 1980. Now no licence is required for setting up a cold storage as far as the union government is concerned. This also means that now no data will be available on the total number of cold storages and their installed capacity in the country in the coming years. According to a report prepared by an expert committee under the chairmanship of Mr. J.L.N. Srivastava, Additional Secretary in the Ministry of Agriculture, the present installed capacity of cold storages in the country for the storage of potatoes is 103 lakh tonnes which needs to be raised to 123 lakh tonnes with an investment of Rs. 650 crores in the next 5 years. This can be done by setting up an additional capacity of 12 lakh tonnes and revamping a capacity of 8 lakh tonnes by rehabilitating some of the closed and sick cold storages. The commitee had suggested setting up of 90 per cent of this additional capacity in the private sector and 10 per cent in the cooperative/ private sector.

INDIA'S IMPORTS OF SOME AGRO PRODUCTS DURING APRIL-DECEMBER, 1998($Million)

Description	Exports in Apr.-Sept.		
	1997	1998	% Growth
Wheat	164.46	199.58	21.35
Fertilizers	973.49	850.22	-12.66
Edible Oils	654.73	1372.96	109.85
Sugar	72.31	194.88	-
Pulses	236.65	66.74	-71.80

The committee has estimated the present installed capacity of the cold storage plants for onions at 4.5 lakh tonnes which needs to be raised by 4.6 lakh tonnes in the coming 5 years.

The committee has also suggested that the lending institutions should provide finance to cold storage plants at a concessional rate of 6 p^r cent, the excise duty on refrigeration plants and sandwich panels used for insulation be abolished. The committee has recommended the replacement of present diffuser cooling system in cold storage plants by gravity/fin cooler system which saves power consumption by 20-25 per cent. Accepting some of the recommendations of the committee, in the 1998-99 Budget, a credit-linked capital subsidy scheme for the construction of cold storages to be implemented by the ministry of agriculture with NABARD has been proposed. There is a plan to create an additional cold storage capacity of 12 lakh tonnes and modernization/upgradation of existing 8 lakh tonnes capacity. An additional capacity of 4.5 takh tonnes for the storage of onions is also proposed to be created.

INSURANCE

In order to provide financial support to farmers in the event of crop failure due to drought, flood, etc and to restore the credit eligibility for the next crop season, a Comprehensive Crop

Insurance Scheme (CCIS) was introduced in April 1985 which covered wheat, paddy, millets, oilseeds and pulses. Under the scheme, half of the premium payable by small and marginal farmers is subsidised equally by the central and the state governments and indemnity claims are shared by the union government and state governments in the ratio of 2:1. The premium is charged at the rate of 2 per cent for rice, wheat and millet and 1 per cent for pulses and oilseeds.

The participation in the scheme is voluntary and the states are free to opt or not to opt for the scheme. The scheme covers only those farmers who avail crop loans from cooperative credit institutions, commercial banks and regional rural banks for producing rice, wheat, millets, oilseeds and pulses.

The sum assured equals to the crop loan disbursed subject to a maximum of Rs. 10,000 per farmer. The scheme also covers indemnity towards loss in yield. If the actual average yield in any area covered by this scheme fails short of the guaranteed yield, the farmer is entitled to an indemnity to the extent of shortfall in yield viz-a-viz the guaranteed yield.

The scheme is being administered by the General Insurance Corporation of India. Since the inception of the scheme in 1985 to Rabi 1996-97 season about 5.82 crore farmers were covered under the scheme and claims of Rs. 1,523 crore were paid against premium collections of mere Rs. 269 crore. Till the end of Rabi 1997-98 season the scheme covered 6.45 crore farmers and claims of Rs. 1,623 crore were paid against a premium collection of only Rs. 313 crores.

The scheme has proved its unviability and the ministry of on some alternate models for agriculture has been working quite some time. In 1998, the union government , formulated a ' Modified Crop Insurance Scheme (MCCIS)' which received an in-principle clearance for introduction from Kharif 1999 by the Cabinet Committee on Economic Affairs in November 1998.

In 1997, an 'Experimental Crop Insurance Scheme (ECIS)' was formulated for implementation in 24 districts for rabi 1997-98 crop which was to cover besides loanee farmers, non-loanee small and marginal farmers also. However, the scheme could be implemented in only 14 districts in 5 States and covered only non-loanee farmers. All the premium payable was subsidized by the central and the state governments in the ratio of 4:1 and the risks and indemnities were also shared by them in 4:1 ratio. About 4.78 lakh farmers were covered for a sum insured of Rs. 172 crores. This scheme also proved a failure as the premium collected was only 2.86 crores and claims paid amounted to Rs. 39.78 crores. From kharif 1998-99, this scheme has been discontinued. In the interim, the CCIS has replaced the ECIS in those districts where the latter is discontinued.

The CCIS is being implemented, at present, in 16 states and union territories. Punjab, Haryana and western Uttar Pradesh have not opted it. The CCIS does not cover commercial crops like sugarcane, cotton, onions, potato, etc. The MCCIS, whose modalities are still being worked out may cover these crops also. There is a proposal to set up a corporation to be called 'Agricultural Insurance Corporation of India' ' as a subsidiary of the General Insurance Corporation to manage its operations.

Under the MMCCIS, the farmers may have to pay premium as against the earlier schemes in which even the premium was being paid by the government. For marginal farmers there will be concessions in premium charges. The minimum support price will take care of the premium charges borne by farmers. To start with the MMCCIS will be introduced in select districts and will cover only foodgrains. Later on it will be extended to all crops including oilseeds, horticulture, etc and all sorts of agricultural activities like fisheries, animal husbandry, sericulture, etc. The scheme will be compulsory for farmers who avail loans from any government institution and optional for others. Since past few years, the union government is also implementing a Livestock Insurance scheme, under which an animal is insured for its present market value or market value at the time of death whichever is lower. In 1997-98, about 16.5 million animals were insured under this scheme . A sum of Rs. 137.40 crore was received as premium and claims of Rs. 93.32 crore were disbursed by GIC which is implementing this scheme.

AGRICULTURAL EXPORTS

The commerce ministry has identified 10 agricultural products for sustainable export promotion. These are : rice, wheat products, coarse grains, spices, cashew, oilmeals, sugar, horticultural products, floriculture products and processed foods. The ministry plans to have a consistent policy for the export of rice, both basmati and non-basmati, wheat products and pulses. The rice meant for export may be exempted from levy obligation. While the imports and exports of rice and pulses can be freed, special efforts may be made to cultivate a sustainable export market for tea, coffee, vegetables and oilmeals. In the past few years, while the exports of some traditional items from India has fallen, that of some new products has started going up.

India's agro exports in 1997-98 are expected to have fallen to Rs. 23,690 crores from Rs. 24,241 crores in 1996-97. In 1997-98, the exports comprised cereals and pulses Rs. 7,520 crores; oilseeds, oils and oilmeals Rs. 4,315 crores; plantation crops (tea, coffee, cashew, tobacco and spices) Rs. 7,292 crores; sugar, molasses, jaggery and alcoholic drinks Rs. 1,695 crores; and fruits, vegetables and other processed foods, Rs. 4,020 crores.

Now there is a worldwide trend towards promoting agricultural exports in almost all the countries where agro-climatic conditions favour cultivation of at least some crops. In fact, the economies of many countries in the world are very much dependent upon agricultural exports in similar lines as that of most Arab countries on the exports of crude and petroleum products. Even in an advanced country like USA which has much more to offer to the world in terms of engineering goods, computers and chemicals, every one out of three acres of land is being dedicated to raise crops for exports.

QUALITY ASPECTS

The European Commission is bringing some more stricter quality legislations on imported food products including animal feeds specially pertaining to permissible levels of aflatoxin, hexane and pesticide residues and other harmful foreign material that enter agricultural produce during post-harvest handling and processing which is expected to affect India's exports,particularly groundnut, cashewnuts, walnuts, pepper, chillies, oilcakes and marine products.

Aflatoxins are very serious problem for many agricultural commodities including maize,chillies,sesame and groundnut. They are potent carcinogenic,immuno-suppressive bio-chemical produced when the aspergillus fungus invades certain agricultural commodities.

The products affected include HPS groundnuts,extractions, spice oils,essential oils,derived from flowers,etc. besides others. The EC intends to limit the levels of hexane in cocoa butter fat and oil to 1 ppm, in defatted protein products and flours to 10 ppm, defatted soya products to 30 ppm and cereal germs to 5 ppm.In a directive if October 27,1997, the EC had reduced the permissible levels of residues in oil extractions of various flowers and spices. The products cover exotic oils derived from jasmine and tuberose, bud oils like cinnamon bark and turmeric root,and seed oils like ajwain, star anise, cardamom, cumin, coriander, fennel, fenugreek, kalongi/nigella, etc.

In HPS groundnuts the aflatoxin level has been brought down to 10ppb in raw nuts and 4 ppb in consumer-ready processed nuts from 10-30 ppb allowed earlier. The levels of aflatoxin in foodstuffs develop mostly when moist unsterilized agricultural products, whether raw or processed, are stored in confined space with improper air ventilation.

The aflatoxin level in HPS groundnut ranges from a low of 10 to over 130-150 and is linked to the quality of groundnuts. The lower count groundnut generally is of better quality as it is picked by workers adept at this job. The EU has set a maximum level of B1 aflatoxin at 5ppb for nuts not meant for direct consumption and at 2 ppb for nuts meant for direct consumption.Even the traders in EU have protested against this as it would increase the price of imported nuts substantially.Earlier India had set the B1 level at 30ppb and many countries including China,USA,Vietnam and Argentina at 20ppb.

The union government is likely to make the implementation of Hazard Analysis and Critical Control Point (HACCP) mandatory for the food sector.It has already asked the Bureau of Indian Standards(BIS) to identify the areas. The BIS is likely to recommend perishable foods,dairy products,marine products,spices,meat products,poultry products and processed food for mandatory implementation of HACCP.The BIS has already prepared and brought out a specification of IS:15000-1998 which is based on HACCP guidelines and recommendations made by the Food Hygiene Committee of FAO/WHO Codex Alimentarius Commission.

RESEARCH AND EXTENSION

The Indian Council of Agricultural Research is an apex institution responsible for the promotion of agricultural research and education in India.Through a network of research institutes, research stations and state level agricultural universities in various parts of the country it has done commendable work in evolving, testing and propagating new farm technologies

ICAR has initiated research on almost all the economically important crops, besides soil sciences, horticulture, animal husbandry, fisheries and agricultural extension. It has played a major role in ushering in the green revolution in the 1970's. Today India is one of the 8 gene centers in the world having three Bureaux dealing with plants, animals and fisheries. ICAR is preparing a 25 year perspective plan to be implemented on a 5-yearly basis with the Master Plan kept 25 years ahead.

However, the country's investment in agricultural research is stagnating at 0.32 per cent of total agricultural GDP for the last two decades. This does not compare well with 0.56 per cent investment being made by 51 developing countries and an average of 2 per cent by the developed countries. The parliamentary standing committee on agriculture has recommended that the investment in agricultural research should be targeted to reach a level of 1 per cent on the total agricultural GDP.

In very recent years, the private sector has also started playing an important role in R&D particularly in the development of high yielding varieties of seeds and propagation of tissue culture plantlets, and in training of farmers by extension work especially the contract farming farmers. The agriculture ministry has set up a national institute of agricultural extension management, 4 extension education training institutes and 15 advanced training institutes in various regions for the training of extension personnel at various levels.

The ICAR has also set up Krishi Vigyan Kedras (KVK's) in the country for the training of farmers, farm women and rural youths. These kendras hold frontline demonstrations of latest developed technologies for farmers and extension workers and also conduct on-farm research to further improve the technology developed to make it more suitable for the agro-climatic conditions in specific regions in which they are located in the country. These kendras work under collaborative participation of scientists, demonstrators, extension workers and farmers.

Since 1974, when the first KVK was established in Pondicherry, 261 KVKs have been established all over the country so far. The KVKs have organized more than 1 lakh training programmes for farmers, In the Ninth Plan period, it is proposed to set up 239 more Krishi Vigyan Kendras. This is a part of the programme to set up 500 KVKs all over the country. There are plans to upgrade 53 research stations, set up under the National Agricultural Research Project, affiliated to various agricultural universities into KVKs.

The ICAR also has plans to set up 80 agricultural technology information centers at agricultural universities and its'research institutes for direct interaction with individual farmers in solving their problems and for supply of literature on modern technology and practices.The ICAR would also cover 104 KVKs under an Agricultural Research Information Services network which will allow ready access to these KVKs all the latest research information on various crops through internet on computers. The union government has identified 26 districts in the country for setting up Agricultural Technology Management Agencies which will coordinate with state agricultural departments, universities and KVKs in providing technical services to farmers.

In the Ninth Plan, there was a proposal to allocate Rs. 2,500 crores for agricultural research and education. The outlay for agriculture education is set to be doubled to Rs. 280 crores during the Ninth Plan and a one-time catch-up grant of Rs. 200 crores is expected to be provided to improve the infrastructure facilities in existing agricultural colleges.

ICAR has earmarked Rs. 2,500 crores for research for developing high-yielding cereal crops and oilseeds through various agricultural research stations in the country. The objective is to double the output of all crops and cereals in the next 10 years.

Since past several years, many of the large fertilizer companies in the public sector like IFFCO, KRIBHCO, NFL, GSFC, RCFL, etc and some companies in the private sectoralso like Nagaduna Fertilizers and]Indo Gulf Fertilizers,have been providing extension education to farmers in cultivation of foodgrains with the object of selling their fertilizers by putting up demonstration plots. Recently some private sector organizations have also started taking interest in extension work. Tata Chemicals and Railis India, for instance, have plans to set up 40 'Tata Kisan Kendras' in the country to disseminate latest information on advancements in seeds, fertilizers, pesticide s and bio-techology to farmers in Uttar Pradesh, Punjab and Haryana.

The VMA Oilseeds Research & Development Institute (VORDI), promoted by the Vanaspati Manufacturers' Association, has also started providing extension education to farmers in North India in the cultivation of oilseeds.

The Defense Food Research Laboratory, Mysore has developed technologies for preparation of shelf-stable foods at ambient temperatures from fruits, vegetables, cereals and meats, though they have not been commercialized so far. The items include ready-to-eat chapaties, pineapple and mango slices, combination preserved meat and dairy products.

The Indian Council of Agricultural Research has proposed to set up a bureau on important micro-organisms for use in agriculture in the Ninth Plan. A National Bureau for Plant Quarantine System has also been proposed to collect, document and preserve genetic resources to face intellectual property rights ([PR) related issues and to strengthen the quarantine and phytosanitation regulations.

The Center for Agriculture & Biosciences International (CABI), London has compiled a database with more than 35 lakh abstracts compiled and abstracted since 1973 from over 100 countries. The software is available on a CD-ROM with ICAR in New Delhi and is available to agricultural universities and 1 research institutes in India on subscription through local area networking and E-mail connections.

It covers data on post harvest and crop improvement techniques which can be of immense help in R&D work. The CABI South Asia collection includes detailed information on pests, diseases and weeds of the world and their national names with synonyms, host range and geographic range. More than 80 per cent of the coverage is from India. The information on the CDROM can be downloaded and printed by subscribers.

The Regional Plant Resource Center, Bhubaneshwar, is setting up a seed bank to preserve over 65,000 varieties of endangered plants and herbs collected from Eastern India in a specially built cold module at -200C and 30 per cent humidity. A large number of plants and herb species have vanished from the forests of Eastern and North-Eastern states. The seeds and tissues can be stored in the bank for 80 to 120 years without any damage.

2

History of Agriculture

Agriculture was developed at least 10,000 years ago, and it has undergone significant developments since the time of the earliest cultivation. Evidence points to the Fertile Crescent of the Middle East as the site of the earliest planned sowing and harvesting of plants that had previously been gathered in the wild. Independent development of agriculture occurred in northern and southern China, Africa's Sahel, New Guinea and several regions of the Americas. Agricultural practices such as irrigation, crop rotation, fertilizers, and pesticides were developed long ago but have made great strides in the past century. The Haber-Bosch method for synthesizing ammonium nitrate represented a major breakthrough and allowed crop yields to overcome previous constraints. In the past century agriculture has been characterized by enhanced productivity, the substitution of labor for synthetic fertilizers and pesticides, selective breeding, mechanization, water pollution, and farm subsidies. In recent years there has been a backlash against the external environmental effects of conventional agriculture, resulting in the organic movement.

ORIGINS

Identifying the exact origin of agriculture remains problematic because the transition from hunter-gatherer societies began thousands of years before the invention of writing. Nonetheless, archaeobotanists/paleoethnobotanists have traced the selection and cultivation of specific food plant characteristics, such as a semi-tough rachis and larger seeds, to just after the Younger Dryas (about 9,500 BC) in the early

Holocene in the Levant region of the Fertile Crescent. There is earlier evidence for use of wild cereals: anthropological and archaeological evidence from sites across Southwest Asia and North Africa indicate use of wild grain (e.g., from the ca. 20,000 BC site of Ohalo II in Israel, many Natufian sites in the Levant and from sites along the Nile in the 10th millennium BC). There is even evidence of planned cultivation and trait selection: grains of rye with domestic traits have been recovered from Epi-Palaeolithic (10,000+ BC) contexts at Abu Hureyra in Syria, but this appears to be a localised phenomenon resulting from cultivation of stands of wild rye, rather than a definitive step towards domestication. It isn't until after 9,500 BC that the eight so-called founder crops of agriculture appear: first emmer and einkorn wheat, then hulled barley, peas, lentils, bitter vetch, chick peas and flax.

These eight crops occur more or less simultaneously on PPNB sites in the Levant, although the consensus is that wheat was the first to be sown and harvested on a significant scale.

Mehrgarh, one of the most important Neolithic (7000 BC to 3200 BC) sites in archaeology, lies on the "Kachi plain of Baluchistan, Pakistan, and is one of the earliest sites with evidence of farming (wheat and barley) and herding (cattle, sheep and goats) in South Asia."

By 7000 BC, sowing and harvesting reached Mesopotamia and there, in the super fertile soil just north of the Persian Gulf, Sumerian ingenuity systematized it and scaled it up. By 6000 BC farming was entrenched on the banks of the Nile River. About this time, agriculture was developed independently in the Far East, probably in China, with rice rather than wheat as the primary crop. Maize was first domesticated, probably from teosinte, in the Americas around 3000-2700 BC, though there is some archaeological evidence of a much older development. The potato, the tomato, the pepper, squash, several varieties of bean, and several other plants were also developed in the New World, as was quite extensive terracing of steep hillsides in much of Andean South America. Agriculture was also independently developed on the island of New Guinea.

Europe By 7000 BC the agriculture were in southern Europe there are evidence of emmer and einkorn wheat, barley, sheep, goats, and pigs suggest that a food producing economy is adopted in Greece and the Agean.

6000-4500 BC Archaeological evidence from various sites on the Iberian peninsula suggest the domestication of plants and animals.

5500 BC Céide Fields in Ireland are the oldest known field systems in the world, this landscape consist of extensive tracks of land enclosed by stone walls.

By 5000 BC the horse were domesticated in the Ukraine,

In China, rice and millet were domesticated by 8000 BC, followed by the beans mung, soy and azuki. In the Sahel region of Africa local rice and sorghum were domestic by 5000 BC. Local crops were domesticated independently in West Africa and possibly in New Guinea and Ethiopia. Evidence of the presence of wheat and some legumes in the 6th millennium BC have been found in the Indus Valley. Oranges were cultivated in the same millennium. The crops grown in the valley around 4000 BC were typically wheat, peas, sesame seed, barley, dates and mangoes. By 3500 BC cotton growing and cotton textiles were quite advanced in the valley. By 3000 BC farming of rice had started. Other monsoon crops of importance of the time was cane sugar. By 2500 BC, rice was an important component of the staple diet in Mohenjodaro near the Arabian Sea. By this time the Indians had large cities with well-stocked granaries. Three regions of the Americas independently domesticated corn, squashes, potato and sunflowers.

THEORY

The reasons for the development of farming may have included climate change, but possibly there were also social reasons (e.g., accumulation of food surplus for competitive gift-giving as in the Pacific Northwest potlatch culture). Most likely there was a gradual transition from hunter-gatherer to agricultural economies after a lengthy period during which some crops were deliberately planted and other foods were gathered in the wild. Although localised climate change is the favoured explanation for the origins of agriculture in the Levant, the fact that farming was 'invented' at least three times elsewhere, and possibly more, suggests that social reasons may have been instrumental.

When major climate change took place after the last ice age c.11,000 BC much of the earth became subject to long dry seasons. These conditions favoured annual plants which die off in the long dry season, leaving a dormant seed or tuber. These plants tended to put more energy into producing seeds than into woody growth. An abundance of readily storable wild grains and pulses enabled hunter-gatherers in some areas to form the first settled villages at this time.

There are several theories as to what drove populations to take up agriculture:

- The Oasis Theory which was original proposed by Raphael Pumpelly in 1908, but popularized by Vere Gordon Childe in 1928 and summarised in his book *Man Makes Himself*[1] This theory maintains that as the climate got drier, communities contracted to oases where they were forced into close association with animals which were then domesticated together with planting of seeds. It has little support now as the climate data for the time does not support the theory.
- The Hilly Flanks hypothesis. Proposed by Robert Braidwood in 1948, it suggests that agriculture began in the hilly flanks of the Taurus and Zagros mountains, and that it developed from intensive focused grain gathering in the region.
- The Feasting model by Bryan Hayden[2] suggests that agriculture was driven by ostentatious displays of power, such as throwing feasts to exert dominance. This required assembling large quantities of food which drove agricultural technology.
- The Demographic theories proposed by Carl Sauer[3] and adapted by Lewis Binford[4] and Kent Flannery. This leads from an increasingly sedentary population, expanding up to the carrying capacity of the local

environment, and requiring more food than can be gathered. Various social and economic factors help drive the need for food.

- The evolutionary/intentionality theory. As proposed by those such as David Rindos[5] the idea that agriculture is an evolutionary adaptation of plants and humans. Starting with domestication by protection of wild plants, followed specialisation of location and then domestication.

ANCIENT AGRICULTURE

By the Bronze Age, wild food contributed a nutritionally insignificant component to the usual diet. If the operative definition of *agriculture* includes large scale intensive cultivation of land, mono-cropping, organized irrigation, and use of a specialized labour force, the title "inventors of agriculture" would fall to the Sumerians, starting ca. 5,500 BC. Intensive farming allows a much greater density of population than can be supported by hunting and gathering, and allows for the accumulation of excess product for off-season use, or to sell/ barter. The ability of farmers to feed large numbers of people whose activities have nothing to do with agriculture was the crucial factor in the rise of standing armies. Sumerian agriculture supported a substantial territorial expansion which along with internecine conflict between cities, made them the first empire builders. Not long after, the Egyptians, powered by farming in the fertile Nile valley, achieved a population density from which enough warriors could be drawn for a territorial expansion more than tripling the Sumerian empire in area.

SUMERIAN AGRICULTURE

In Sumer, barley was the main crop, but wheat, flax, dates, apples, plums, and grapes were grown as well. Mesopotamia was blessed with flooding from the Tigris and Euphrates rivers but floods came in late spring or early summer from snow melting from the Turkish mountains. With Salt deposits under the soil, all of this made Mesopotamia very hard to farm.[6] The earliest known sheep and goats were also domesticated and were in a much larger quantity than cattle. Sheep were mainly kept for meat and milk, and butter and cheese were made from the latter. Ur, a large town that covered about 50 acres (20 hectares), had 10,000 animals kept in sheepfolds and stables and 3,000 slaughtered every year. The city's population of 6,000 included a labour force of 2,500 cultivated 3,000 acres (12 km^2) of land. The labour force contained storehouse recorders, work foremen, overseers, and harvest supervisors to supplement labourers. Agricultural produce was given to temple personnel, important people in the community, and small farmers.

The land was plowed by teams of oxen pulling light unwheeled plows and grain was harvested with sickles in the spring. Wagons had solid wheels covered by leather tires kept in position by copper nails and were drawn by oxen and the Syrian onager (now extinct). Animals were harnessed by collars, yokes, and headstalls. They were controlled by reins, and a ring through the nose or upper lip and a strap under the jaw. As many as four animals could pull a wagon at one time. Though some hypothesize that Domestication of the horse occurred as early as 4000 BC in the Ukraine, the horse was definitely in use by the Sumerians around 2000 BC.

Chinese agriculture

The unique tradition of Chinese agriculture has been traced to the pre-historic Xianrendong Relics and Diaotonghuan Relics (c. 12 0000 BC-7500 BC). Chinese historical and governmental records of the Warring States (481 BC-221 BC), Qin Dynasty (221 BC-207 BC), and Han Dynasty (202 BC-220 AD) eras allude to the use of complex agricultural practices, such as a nationwide granary system and widespread use of sericulture. However, the oldest extant Chinese book on agriculture is the *Chimin Yaoshu* of 535 AD, written by Jia Sixia. Although much of the literature of the time was elaborate, flowery, and allusive, Jia's writing style was very straightforward and lucid, a literary approach to agriculture that later Chinese agronomists after Jia would follow, such as Wang Zhen and his groundbreaking *Nong Shu* of 1313 AD. Jia's book was also incredibly long, with over one hundred thousand written Chinese characters, and quoted 160 other Chinese books that were written previously (but no longer survive). The contents of Jia's 6th century book include sections on land preparation, seeding, cultivation, orchard management, forestry, and animal husbandry. The book also includes peripherally related content covering trade and culinary uses for crops.

For agricultural purposes, the Chinese had innovated the hydraulic-powered trip hammer by the 1st century BC. Although it found other purposes, its main function to pound, decorticate, and polish grain that otherwise would have been done manually. The Chinese also innovated the square-pallet chain pump by the 1st century AD, powered by a waterwheel or an oxen pulling a on a system of mechanical wheels. Although the chain pump found use in public works of providing water for urban and palatial pipe systems,[12] it was used largely to lift water from a lower to higher elevation in filling irrigation canals and channels for farmland.

Papuan Agriculture

Ancient Papuans are thought to have begun practicing agriculture around 7000 BC. They began domesticating sugarcane and root crops. Pigs may also have been domesticated around this time. By 3000 BC, Papuan agriculture was characterised by water control for irrigation.

Indian Agriculture

Evidence of the presence of wheat and some legumes in the 6th millennium BC have been found in the Indus Valley. Oranges were cultivated in the same millennium. The crops grown in the valley around 4000 BC were typically wheat, peas, sesame seed, barley, dates and mangoes. By 3500 BC cotton growing and cotton textiles were quite advanced in the valley. By 3000 BC farming of rice had started. Other monsoon crops of importance of the time was cane sugar. By 2500 BC, rice was an important component of the staple diet in Mohenjodaro near the Arabian Sea.

The Indus Plain had rich alluvial deposits which came down the Indus River in annual floods. This helped sustain farming that formed basis of the Indus Valley Civilization at Harappa. The people built dams and drainage systems for the crops. By 2000 BC tea, bananas and apples were being cultivated in India. There was coconut trade with East Africa in 200 BC. By 500 AD, eggplants were being cultivated.

Roman Agriculture

Roman agriculture built off techniques pioneered by the Sumerians, with a specific emphasis on the cultivation of crops for trade and export. Romans laid the groundwork for the manorial economic system, involving serfdom, which flourished in the Middle Ages.

Mesoamerican Agriculture

In Mesoamerica, the Aztecs were some of the most innovative farmers of the ancient world and farming provided the entire basis of their economy. The land around Lake Texcoco was fertile but not large enough to produce the amount of food needed for the population of their expanding empire. The Aztecs developed irrigation systems, formed terraced hillsides, and fertilized their soil. However, their greatest agricultural technique was the *chinampa* or artificial islands also known as "floating gardens". These were used to make the swampy areas around the lake suitable for farming. To make chinampas, canals were dug through the marshy islands and shores, then mud was heaped on huge mats made of woven reeds. The mats were anchored by tying them to posts driven into the lake bed and then planting trees at their corners that took root and secured the artificial islands permanently. The Aztecs grew corn, squash, vegetables, and flowers on chinampas.

Andean Agriculture

The Andean civilizations were predominantly agricultural societies; the Incas took advantage of the ground, conquering the adversities like the Andean area and the inclemencies of the weather. The adaptation of agricultural technologies that already were used previously, allowed the Incas to organize the production a diversity of products of the coast, mountain and jungle, so them could be able to redistribute to villages that did not have access to other regions. The technological achievements reached to agricultural level, had not been possible without the workforce that was at the disposal of the Sapa Inca, as well as the road system that was allowing to store adequately the harvested resources and to distribute them for all the territory.

Muslim Agricultural Revolution

From the 8th century, the medieval Islamic world witnessed a fundamental transformation in agriculture known as the "Muslim Agricultural Revolution", "Arab Agricultural Revolution", or "Green Revolution". Due to the global economy established by Muslim traders across the Old World during the "Afro-Asiatic age of discovery" or "Pax Islamica", this enabled the diffusion of many crops, plants and farming techniques between different parts of the Islamic world, as well as the adaptation of crops, plants and techniques from beyond the Islamic world, distributed throughout Islamic lands which normally would not be able to grow these crops. These techniques included crop rotation, irrigation and pest control. Some have referred to the diffusion of numerous crops during this period as the "Globalisation of Crops",which, along with increased mechanization of agriculture, led to major changes in economy, population distribution, vegetation cover,[18] agricultural production and income, population levels, urban growth, the distribution of the labour force, linked industries, cooking and diet, clothing, and numerous other aspects of life in the Islamic world.

Serfdom became widespread in eastern Europe in the Middle Ages. Medieval Europe owed much of its development to advances made in Islamic areas, which flourished culturally and materially while Europe and other Roman and Byzantine administered lands entered an extended period of social and economic stagnation. As early as the ninth century, an essentially modern agricultural system became central to economic life and organization in the Arab caliphates, replacing the largely export driven Roman model. The great cities of the Near East, North Africa and Moorish Spain were supported by elaborate agricultural systems which included extensive irrigation based on knowledge of hydraulic and hydrostatic principles, some of which were continued from Roman times. In later centuries, Persian Muslims began to function as a conduit, transmitting cultural elements, including advanced agricultural techniques, into Turkic lands and western India. The Muslims introduced what was to become an agricultural revolution based on four key areas:

- Development of a sophisticated system of irrigation using machines such as norias, water mills, water raising machines, dams and reservoirs. With such technology they managed to greatly expand the exploitable land area.
- The adoption of a scientific approach[19] to farming enabled them to improve farming techniques derived from the collection and collation of relevant information throughout the whole of the known world.[19] Farming manuals were produced in every corner of the Muslim world detailing where, when and how to plant and grow various crops. Advanced scientific techniques allowed leaders like Ibn al-Baytar to introduce new crops and breeds and strains of livestock into areas where they were previously unknown.
- Incentives based on a new approach to land ownership and labourers' rights, combining the recognition of private ownership and the rewarding of cultivators with a harvest share commensurate with their efforts. Their counterparts in Europe struggled under a feudal system in which they were almost slaves (serfs) with little hope of improving their lot by hard work.
- The introduction of new crops transforming private farming into a new global industry exported everywhere,[16] including Europe, where farming was mostly restricted to wheat strains obtained much earlier via central Asia. Spain received what she in turn transmitted to the rest of Europe; many agricultural and fruit-growing processes, together with many new plants, fruit and vegetables. These new crops included sugar cane, rice, citrus fruit, apricots, cotton, artichokes, aubergines, and saffron. Others, previously known, were further developed. Muslims also brought to that country lemons, oranges, cotton, almonds, figs and sub-tropical crops such as bananas and sugar cane. Several were

later exported from Spanish coastal areas to the Spanish colonies in the New World. Also transmitted via Muslim influence, a silk industry flourished, flax was cultivated and linen exported, and esparto grass, which grew wild in the more arid parts, was collected and turned into various articles.

Agriculture in the Middle Ages

During the Middle Ages, Muslim farmers in North Africa and the Near East developed and disseminated agricultural technologies including irrigation systems based on hydraulic and hydrostatic principles, the use of machines such as norias, and the use of water raising machines, dams, and reservoirs. They also wrote location-specific farming manuals, and were instrumental in the wider adoption of crops including sugar cane, rice, citrus fruit, apricots, cotton, artichokes, aubergines, and saffron. Muslims also brought lemons, oranges, cotton, almonds, figs and sub-tropical crops such as bananas to Spain.

Renaissance Agriculture

The invention of a three field system of crop rotation during the Middle Ages, and the importation of the Chinese-invented moldboard plow, vastly improved agricultural efficiency.

After 1492 the world's agricultural patterns were shuffled in the widespread exchange of plants and animals known as the Columbian Exchange. Crops and animals that were previously only known in the Old World were now transplanted to the New and vice versa. Perhaps most notably, the tomato became a favorite in European cuisine, and maize and potatoes were widely adopted. Other transplanted crops include pineapple, cocoa, and tobacco. In the other direction, several wheat strains quickly took to western hemisphere soils and became a dietary staple even for native North, Central and South Americans.

Agriculture was a key element in the Atlantic slave trade, Triangular trade, and the expansion by European powers into the Americas. In the expanding Plantation economy, large plantations producing crops including sugar, cotton, and indigo, were heavily dependent upon slave labor.

British Agricultural Revolution

Between the 16th century and the mid-19th century, Great Britain saw a massive increase in agricultural productivity and net output. New agricultural practices like enclosure, mechanization, four-field crop rotation and selective breeding enabled an unprecedented population growth, freeing up a significant percentage of the workforce, and thereby helped drive the Industrial Revolution.

By the early 1800s, agricultural practices, particularly careful selection of hardy strains and cultivars, had so improved that yield per land unit was many times that seen in the Middle Ages and before.

The 18th and 19th century also saw the development of glasshouses, or greenhouses, initially for the protection and cultivation of exotic plants imported to Europe and North America from the tropics.

Experiments on Plant Hybridization in the late 1800s yielded advances in the understanding of plant genetics, and subsequently, the development of hybrid crops.

Increasing dependence upon monoculture crops lead to famines and food shortages, most notably the Irish Potato Famine (1845–1849).

Storage silos and grain elevators appeared in the 19th century.

RECENT HISTORY

New Technologies

With the rapid rise of mechanization in the late 19th and 20th centuries, particularly in the form of the tractor, farming tasks could be done with a speed and on a scale previously impossible. These advances, joined to science-driven innovations in methods and resources, have led to efficiencies enabling certain modern farms in the United States, Argentina, Israel, Germany and a few other nations to output volumes of high quality produce per land unit at what may be the practical limit. The development of rail and highway networks and the increasing use of container shipping and refrigeration in developed nations have also been essential to the growth of mechanized agriculture, allowing for the economical long distance shipping of produce.

While chemical fertilizer and pesticide have existed since the 19th century, their use grew significantly in the early twentieth century. In the 1960s, the Green Revolution applied western advances in fertilizer and pesticide use to farms worldwide, with varying success.

Other applications of scientific research since 1950 in agriculture include gene manipulation, and Hydroponics.

New Criticisms

Though the intensive farming practices pioneered and extended in recent history generally led to increased outputs, they have also led to the destruction of farmland, most notably in the dust bowl area of the United States following World War I.

As global population increases, agriculture continues to replace natural ecosystems with monoculture crops.

In the past few decades, western consumers have become increasingly aware of, and in some cases critical of, widely used intensive agriculture practices, contributing to a rise in popularity of organic farming, the growth of the Slow Food movement, and an ongoing discussion surrounding the potential for sustainable agriculture.

3

Agriculture in India

India's record of progress in agriculture over the past four decades has been quite impressive. The agriculture sector has been successful in keeping pace with rising demand for food. The contribution of increased land area under agricultural production has declined over time and increases in production in the past two decades have been almost entirely due to increased productivity. Contribution of agricultural growth to overall progress has been widespread. Increased productivity has helped to feed the poor, enhanced farm income and provided opportunities for both direct and indirect employment. The success of India's agriculture is attributed to a series of steps that led to availability of farm technologies which brought about dramatic increases in productivity in 70s and 80s often described as the **Green Revolution** era. The major sources of agricultural growth during this period were the spread of modern crop varieties, intensification of input use and investments leading to expansion in the irrigated area. In areas where 'Green Revolution' technologies had major impact, growth has now slowed.

New technologies are needed to push out yield frontiers, utilize inputs more efficiently and diversify to more sustainable and higher value cropping patterns. At the same time there is urgency to better exploit potential of rainfed and other less endowed areas if we are to meet targets of agricultural growth and poverty alleviation. Given the wide range of agroecological setting and producers, Indian agriculture is faced with a great diversity of needs, opportunities and prospects. Future growth needs to be more rapid, more widely distributed and better targeted. These challenges have profound implications for the way farmers' problems are conceived, researched and transferred to the farmers. On the one hand agricultural research will increasingly be required to address location specific problems facing the communities on the other the systems will have to position themselves in an increasingly competitive environment to generate and adopt cutting edge technologies to bear upon the solutions facing a vast majority of resource poor farmers.

In the past agriculture has played and will continue to play a dominant role in the growth of Indian economy in the foreseeable future. It represents the largest sector producing around 28 percent of the GDP, is the largest employer providing more than 60 percent of the jobs and is the prime arbiter of living standards for seventy percent of India's population living in the rural areas. These factors together with a strong determination to achieve self-sufficiency in food grains production have ensured a high priority for agriculture sector in the successive development plans of the country.

An important facet of progress in agriculture is its success in eradication of its critical dependence on imported foodgrains. In the 1950's nearly 5 percent of the total foodgrains available in the country were imported. This dependence worsened during the 1960's when two severe drought years led to a sharp increase in import of foodgrains. During 1966 India had to import more than 10 million tonnes of foodgrains as against a domestic production of 72 million tonnes. In the following year again, nearly twelve million tonnes had to be imported. On the average well over seven percent of the total availability of foodgrains during the 1960s had to be imported.

Indian agriculture has progressed a long way from an era of frequent droughts and vulnerability to food shortages to becoming a significant exporter of agricultural commodities. This has been possible due to persistent efforts at harnessing the potential of land and water resources for agricultural purposes. Indian agriculture, which grew at the rate of about 1 percent per annum during the fifty years before independence, has grown at the rate of about 3 percent per annum in the post independence era.

AGRICULTURE - SUB-SECTORS

Indian agriculture broadly consists of four sub-sectors. Agriculture proper including all food-crops oilseeds, fiber, plantation crops, fruits and vegetables is the largest accounting for nearly 70 percent of the agriculture sector as a whole. The rapid growth in this sub-sector through exploitation of wastelands and fallows, spread of irrigation and adoption of production enhancing technologies was critical in transforming India from a country vulnerable to food shortages to one of exportable surplus. Although this sub-sector has made impressive progress its share in the sector as a whole has declined from 78 percent in 1960-61 to less than 70 percent by early 90s.

Table. Agriculture sub-sectors – value of output (%)

Sub-sector	1960-61	1970-71	1980-81	1993-94
Agriculture	78.0	79.3	75.3	68.5
Livestock	16.6	15.1	17.2	24.5
Forestry	4.4	4.4	5.9	3.8
Fisheries	1.0	1.2	1.6	3.2
	100.0	100.0	100.0	100.0
Total value at Current price (Rs. Billion)	89	218	615	2740

Correspondingly the share of livestock sector has increased considerably. The livestock industry has grown from Rs. 15 billion in early 1960s to Rs. 100 billion by 1980-81 and Rs. 672 billion by 1993-94. In nominal terms the sector grew at almost 15 percent per annum during 1980s. Milk production, which was almost stagnant for two decades ending 1970, grew by over 5 percent per annum in the 80s. Similarly, production of eggs increased at the rate of about 6.5 percent during the same period. As a result the share of livestock increased from about 17 percent till early 80s to 25 percent by 1993-94.

Though it plays relatively a minor role within the sector as a whole, fishing sub-sector activities have been on the rise. The sub-sector has grown from only Rs. 3 billion in 1970-71 to nearly Rs. 90 billion in 1993-94. The growth was particularly rapid in 70s and 80s. Value added increased at over 5 percent per annum during this period.

In real terms forestry and logging activities have been on the decline since mid seventies. As of 1993-94, the size of the industry in terms of value of output was 103 billion.

Over the past three decades, the country has successfully transformed itself from a food deficit economy to one which is essentially self sufficient in availability of foodgrains and other essential commodities, albeit only at the prevailing level of effective demand. Annual aggregate foodgrains production, which averaged about 82 million tonnes in 1960-61 increased to 123.7 and 172.5 million tonnes for the trienniums ending 1980-81 and 1990-91 respectively. Current (1998-99) production level is 195 million tonnes and the country has been able to accumulate substantial, (35 million tonnes) stocks of foodgrains to cope up with any sudden difficulties arising from drought or a similar situation in any part of the country.

Table. Foodgrains Production

Year	Area (million ha)	Production (million tonnes)	Yield kg/ha
1950-51	97.32	50.82	522
1960-61	115.58	82.02	709
1970-71	124.32	108.42	872
1980-81	126.67	129.59	1023
1990-91	127.84	176.39	1379
1998-99	124.0	195.25	1574

Increased outputs, have been achieved chiefly by adopting, since mid sixties, a strategy aimed at increasing foodgrains production by concentrating public sector efforts and resources in regions with a high potential for quick and substantial productivity gains through increased cropping intensity and average yields. These were the areas favoured by agroclimatic resource conditions and where irrigation facilities already existed or could be developed relatively rapidly. The main elements of this strategy were:

(i) expansion of irrigation coverage,

(ii) increased provision and utilization of key inputs – mainly high yielding varieties (HYVs) of crops, mainly of wheat and rice and chemical fertilizers and plant protection chemicals,

(iii) expansion and improvement of institutional support services such as research and extension and

(iv) price policies favourable to producers of major foodgrains.

The success of this strategy was made possible by development and availability of replicable production technology packages, so called 'Green Revolution' technologies. Irrigation facilitated double cropping and widespread adoption of HYVs. The HYVs performed particularly well under irrigated conditions, were highly responsive to fertilizers and their short duration permitted increases in cropping intensities.

Irrigation development was the cornerstone of the strategy. Undivided India was amongst the largest irrigated areas in the world. With partition nearly one-third of the irrigated area went to Pakistan. At the time of independence the net irrigated area was 20.9 million ha (gross irrigated area 22.6 million ha). Recognizing large-scale development of irrigation facilities as critical to rapid agricultural growth, the country has spent about Rs. 45,000 crores on irrigation development in the first four decades after independence. During the period 1950-51 to 1965-66 development of irrigation through government canals grew from 7.2 million ha to 9.8 million ha – a growth rate of 2.1 percent per annum. During 1970s this pace dropped slightly to 1.9 percent. In 1980s the rate of increase dropped significantly to 1.1 percent per annum.

The growth of tube-well irrigation, however, increased rapidly from 4.5 million ha in 1970-71 to 9.5 million ha in 1980-81 and then to 14.3 million ha by 1990-91. The net irrigated area increased from 31 million ha in 1970-71 to 53.5 million ha in 1995-96 which corresponds to 22 percent of the net sown area in 1970-71 and 37.63 percent in 1995-96 *(Tables 3, 4)*. With improvements in irrigation efficiency the gross irrigated areas has increased to 71.51 million ha. The percentage of gross cropped area service by irrigation increased from 18.3 percent in 1960-61 to 23.0 percent in 1970-71 and to over 38 percent at present.

Table. India – irrigated area

Year	Net	Gross	Net as % of net sown	Gross as % of gross cropped
	million ha		%	
1970-71	31.10	38.19	22.17	23.04
1980-81	38.72	49.78	27.66	28.83
1990-91	48.02	63.20	33.61	34.03
1995-96	53.51	71.51	37.63	38.33

Table. Sources of irrigation

Year	Net Irrigated Area (million ha)	Sources				
		Canals	Tanks	Tube-wells	Other wells	Other Sources
		(% net irrigated area)				
1960-61	24.80	10.4	4.6	0.2	7.2	2.4
1970-71	31.10	41.28	13.22	14.34	23.88	7.29
1980-81	38.72	39.49	8.22	24.62	21.08	6.59
1990-91	48.02	36.34	6.13	29.69	21.73	6.11
1995-96	53.01	32.04	5.81	33.52	22.16	6.47

Fertilizers have constituted yet another key input in addition to expanded irrigation and spread of HYVs in achieving goals of high production and productivity. India currently occupies third position in the world, after China and USA, in terms of fertilizer production and consumption. Consumption of fertilizers has increased from 1.54 million tonnes in 1967-68, representing the pre green revolution era, to 17.31 million tonnes currently (1997-98).

The average per hectare use of fertilizers currently around 85 kg per hectares is the lowest among several Asian countries. However, rice and wheat account for a major fraction, around 65 percent of the total fertilizer consumed in the country, with very little fertilizers going to the rainfed areas. According to some current projections, fertilizer's use will need to increase to 30-35 million tonnes to meet the foodgrains need of 2020. The demand for nutrients will stretch by almost another 15 million tonnes if requirements for horticulture, vegetables and plantation and commercial crops are included. At present domestic production of N and P fertilizers (13.42 million tonnes) falls short of consumption by over 20 percent. In addition the entire requirement of K fertilizer is imported.

Most agricultural development programmes initiated in 1960s were concentrated in regions of high potential. Thus five states, Punjab, Haryana, Uttar Pradesh, Andhra Pradesh and Tamil Nadu account for 50 percent of the country's net irrigated and 53 percent of the gross irrigated area. The combination of expanding irrigation coverage and widespread adoption of short duration HYVs led to significant increases in cropping intensities. Acreage cropped more than once per year increased from 13 million ha in 1950-51 to about 44 million ha at present. Average cropping intensity for the country as a whole rose from 115 percent in 1960-61 to 131 in 1993-94. By 1993-94 cropping intensity has risen to 187 percent in Punjab, 167 percent in Haryana and 142 percent in Uttar Pradesh.

Table. Cropping intensity

Years	Net sown area	Gross cropped area	Area cropped more than once	Cropping intensity
	million ha			(%)
1950-51	118.8	131.9	13.1	111.1
1960-61	133.2	152.8	19.6	114.3
1970-71	140.3	165.8	25.5	118.2
1980-81	140.0	172.6	32.6	123.3
1990-91	143.0	185.7	42.7	129.9
1993-94	142.0	186.4	44.4	131.2

An important consequence of the strategies adopted since sixties has been to boost production of, chiefly, two crops rice and wheat. Their share in total foodgrains production went up from 57 percent in 1970-71 to more than 75 percent in 1990-

91. Production of foodgrains other than rice and wheat did not increase significantly and in the eastern region even the yield of rice did not increase. Agricultural production and income rose substantially in the north-western states of Punjab, Haryana, Western Uttar Pradesh, parts of Rajesthan, Tamil Nadu and Andhra Pradesh. By contrast productivity and output growth have been modest in eastern and central India and in deccan plateau. Progress was particularly slow in rainfed areas, which account for over 60 percent of the cropped area and where a great majority of rural poor are concentrated. An important impact of the strategies pursued in the 'Green Revolution' period has been intensification of regional disparities and imbalances in agricultural development and food availability and hence levels of food security. For the country as a whole while per capita availability of cereals has increased substantially, that of pulses has decreased significantly.

Table. Rice and Wheat Production

Production (million tonnes)

Year	Rice	Wheat	Foodgrains	R+W as % of Foodgrains
	million tonnes			
1970-71	40.80	20.86	108.4	57.4
1980-81	49.91	34.55	129.6	65.4
1990-91	72.78	53.03	176.4	75.0
1996-97	81.74	69.35	199.4	75.0

In summary an annual increase in foodgrains production of 3.22 percent during fifties was mainly because of expansion in area. Sixties recorded a low annual growth rate of 1.72 percent necessitating large-scale imports of foodgrains. Annual growth of 2.08 percent was recorded during seventies. This decade was the turning point in India's foodgrains economy leading to self-sufficiency through significant productivity increase first in wheat and later in rice in the eighties. An annual growth of 3.5 percent in foodgrains in eighties was the hallmark of green revolution that enabled India to become self-sufficient and even a marginal exporter. The pace of growth slowed in nineties barely making or even slower than the population growth rate. This is a matter of concern.

Table. Annual compound growth rate of foodgrains production

Crop	1950-51 to 1959-60	1960-61 to 1969-70	1970-71 to 1979-80	1980-81 to 1989-90	1990-91 to 1997-98
Rice	3.28	- 8.05	1.91	4.29	1.53
Wheat	4.51	5.90	4.69	4.24	3.67
Coarse cereals	2.75	1.48	0.74	0.74	- 0.49
Total cereals	3.00	2.51	2.37	3.63	1.84
Pulses	2.72	1.35	- 0.54	2.78	0.76
Total foodgrains	3.22	1.72	2.08	3.54	1.66

Table. Per capita net availability of cereals and pulses per day (g)

Year	Cereals	Pulses	Total
1951	334	61	395
1961	400	69	469
1971	418	51	469
1981	417	38	455
1991	468	42	510
1998	451	33	484

HORTICULTURE

The diversity of physiographic, climate and soil characteristics enables India to grow a large variety of horticultural crops – fruits, vegetables, flowers, spices, aromatic and medicinal plants, plantation crops etc. India is the largest producer of fruits in the world and second largest producer of vegetables. The area under fruits estimated at 1.45 million ha in 1970-71 grew to 2.8 million ha in 1991-92 and then more rapidly to 5 to 6 million ha by 1994-95. This sector is likely to grow rapidly in the future both on account of internal demands and export opportunities.

ANIMAL HUSBANDRY AND FISHERIES

Animal husbandry and dairying sub-sector plays an important role in overall economy and in social development. The contribution of the sub-sector is estimated to be about 25 percent of the total value of output of agricultural sector.

Table. Production of milk, eggs and fish in India

Year	Milk production / availability million tonnes/annum	g./day	Fish (000 tonnes)	Eggs (million)
1950-51	17.0	124	752	1832
1960-61	20.0	124	1160	2881
1970-71	22.0	112	1756	6172
1980-81	31.6	128	2442	10060
1990-91	53.9	176	3836	21101
1998-99	75.0	210	5388	28400

The sector also plays a significant role in supplementing family incomes and generating employment in the rural sector particularly among the land-less, small and marginal farmers and women besides providing nutritious food. Production of suitable cross breeds and their wider adoptions has contributed to increasing country's milk production, which has now reached 75 million tonnes annually. Similarly, genetic improvement and better management practices have markedly pushed up production of poultry and eggs.

Table. Number of holdings (million)

Category of Farmers	1970-71	1980-81	1990-91
Marginal	35.7 (50.6)*	50.1 (56.4)	62.1 (59.0)
Small	13.4 (19.0)	16.1 (18.1)	20.0 (19.0)
Semi-medium	10.7 (15.2)	12.5 (14.0)	13.9 (13.2)
Medium	7.9 (11.2)	8.1 (9.1)	7.6 (7.3)
Large	2.8 (3.9)	2.2 (2.4)	1.7 (1.6)
All groups	70.5 (100)	89.0 (100)	105.3 (100)

*Figures in parenthesis represent the percentage of holdings.

Through the overall contribution of fisheries sub-sector is small, development of prolific and fast growing forms of several common fish species coupled with breakthrough in breeding under captivity, fish seed production and multi-layer fish culture has resulted in registering a very high annual growth of 11 percent in aquaculture production during the past decade.

Table. Area operated - % of total area

Category of Farmers	1970-71	1980-81	1990-91
Marginal	9.0	12.1	14.9
Small	11.9	14.2	17.3
Semi-medium	18.5	21.1	23.1
Medium	29.7	29.6	27.2
Large	30.9	23.0	17.4
All groups (million ha)	**162.1 (100)**	**163.8 (100)**	**165.6 (100)**

SUSTAINABILITY CONCERNS

Several indicators highlight increasing concerns of sustainability in areas which have largely contributed to increased production in the 'Green Revolution' era. Adoption of high yielding cultivators is virtually complete. Almost entire wheat and rice crops in the states of Punjab, Haryana and Western Uttar Pradesh are irrigated. In the higher production regions yields are plateauing and most traditional sources of productivity growth having been exhausted future gains in production have to come from elsewhere.

At farmers' level concerns are being expressed in several ways. Many farmers believe that the input levels have to be continuously increased in order to maintain high yields. In sixties and seventies most farmers used only nitrogenous and phosphate fertilizers to achieve high yields. Due to widespread deficiencies of several secondary and micronutrients, most farmers now have to apply higher doses and a greater variety of fertilizers to maintain crop yields.

Results from many long term studies on rice-wheat cropping system show a declining yield trend when input levels were kept constant – thus the growth rate of system productivity has been declining relative to growth rate of nutrients use. Lowering of groundwater tables due to intensive rice-wheat system in many areas is resulting in increased costs of lifting water in the intensively cultivated high production areas, diseases and pest problems are turning more serious than ever before and pose both short and long large problems. It is reported that some weeds have developed resistance to the commonly used herbicides. What this implies is that the farmers are applying increasing amount of herbicide incurring increasing cost without the benefit of effective control. Pesticide residues entering the food chain and overall safety in use of pesticides continue to be serious problems.

Other emerging problems threatening sustainability of intensive cropping system e.g. rice-wheat include loss in biodiversity related issues. Large areas planted to a single/few varieties of a crop is a potential cause of concern. As the diversity is reduced natural processes that control and affect habitat quality and genetic expression weaken and for this reason internal and natural control mechanisms must be replaced by more externally applied artificial controls in the form of management and inputs which in due course lead the system towards unsustainability.

Groundwater is the major source of meeting the irrigation needs of irrigated agriculture. Currently about half the area under irrigation in the country is irrigated from groundwater sources. Large-scale groundwater development has led to fall in the water table in many areas. Over pumping is leading to declining water table levels and failure of tube-wells. Pumping costs are increasing, as is the energy consumption. In the coastal areas this has led to ingress of sea water, with serious environmental implications.

Changes in water quality are adversely affecting agriculture and vice-versa. Inefficient and/or over use of fertilizers and pesticides in agriculture and untreated disposal of industrial and urban wastes are leading to increasing contamination by such elements as lead, zinc, copper, chromium, cadmium particularly in areas having high industrial activity e.g. in districts of Ludhiana, Faridabad, Kanpur, Varanasi etc.

An increase in the content of arsenic has been reported in several of the districts of West Bengal. This is attributed amongst other causes to the lowering of groundwater table due to excessive groundwater withdrawal and is leading to serious and widespread toxicity problems adversely affect the health of hundreds of thousands people of the region.

CHANGING LAND-USE AND FUTURE OF AGRICULTURE

One of the most important consequences of growing pressure on land is the declining trend in the average farm size and the pattern of holdings. According to the latest Agricultural Census in 1970-71 there were 70 million holdings operating 162 million ha. By 1990-91 there were 105 million holdings operating 165 million ha. The average farm size decreased from 2.30 ha in 1970-71 to 1.57 ha in 1990-91. As of 1990-91 about 78 percent of holdings were small (1.0 to 2.0 ha) and marginal (<1.0 ha). A little more than 20 percent of the farmers were semi-medium (2.0 to 4.0 ha) and medium (4.0 to 10.0 ha).

Large farmers (>10.0 ha) constituted only 1.6 percent of the total holdings. Over the twenty-year period since 1970 the proportion of marginal farmer has increased from 50 to 59 percent and that of large farmer has declined from about 4 to 1.6 percent.

Table. India – Land Use Categories (million ha)

Classification	1960-61	1970-71	1980-81	1990-91	1993-94
Reporting area	298.5	305.0	305.0	305.0	305.0
Area under non-agricultural uses	14.8	16.5	19.7	21.1	22.0
Barren and un-cultivable	35.9	28.2	20.0	19.4	19.0
Cultivable waste	19.2	17.5	16.7	15.0	14.5
Old fallow	11.2	8.8	9.9	9.7	9.7
Forests	54.1	63.9	67.5	67.8	68.4
Permanent pastures	14.0	13.3	12.0	11.8	11.2
Land under misc. trees etc.	4.5	4.3	3.6	3.8	3.7
Net sown area	133.2	140.3	140.0	143.0	142.1
Current fallow	11.6	11.1	14.8	13.7	14.3

The proportion of total area operated by marginal farmers increased from nine percent in 1970-71 to nearly 15 percent in 1990-91 while the proportion of large farmers declined from about 31 percent to 17 percent in the same period. The size of average holding is very unevenly distributed among the states. States with relatively large average size of operational holding are a mixed lot – they include states with large tracts of barren lands e.g. Rajasthan, Maharashtra and Madhya Pradesh on the one hand and agriculturally advanced state like Punjab on the other. These trends in farm size changes will have a profound effect on the future agricultural development strategies.

Table. Livestock and poultry population (million)

Category	1982	1987	1992
Cattle	192.45	199.69	204.53
Buffaloes	69.78	75.96	83.49
Sheep	48.76	45.70	50.79
Goats	95.25	110.20	115.28
Total (Livestock)	419.59	445.28	470.15
Total Poultry	207.74	275.32	307.10

Although India's population growth rate has slowed from 2.1 percent in 1980s to 1.8 percent in the 1990s and is expected to slow further in the coming decades, yet the population is projected to reach 1.33 billion by 2020 from the current one billion. The urban share of total population is projected to increase from 26 percent to 35 percent of the total population. Although incidence of poverty is falling, it is estimated that in 93-94 (upto which data is available) 320 million people constituting 36 percent of the population were below the officially defined poverty line.

The nature of the poverty line has been shifting. About 30 years ago 48.4 percent of those living in rural areas were poor and 20 percent of those living in the urban areas were classed as poor. Recent studies show that the number of poor in urban areas have been increasing at relatively higher rate compared to the rural areas. At present those below the poverty line in rural sector constitute 37 percent of the population while in the urban sector the percentage is 32 percent. In the context of poverty alleviation, therefore, emphasis will be required to be placed both on production of food by the poor as well as on the availability of food for the urban poor.

It needs to be recognized that a large proportion of the rural poor are located in regions of low potential for food production e.g. arid and semi-arid areas, hilly regions, degraded land and forest areas. Widespread hunger and malnutrition are the direct manifestation of poverty and will call for increasing efforts to produce more food at affordable price.

Increasing population and economic growth are changing patterns of land use making potentially unsustainable demands on the country's natural resources.

Since early fifties the net area sown was expanded rapidly at first but at a diminishing rate since 1970 to reach approximately 142 million ha at present.

During 1950s and 1960s areas under agriculture expanded substantially as the fallows were reduced and cultivable wastes were put under the plough.

The net area sown increased from 119 million ha in 1950-51 to 133 million ha by 1960-61 and further to 140 million ha by 1970-71. Fallow lands declined from 28 million ha in 1950-51 to 20 million ha by 1970-71. Cultivable wastelands declined from 23 to 17.5 million ha.

Land use intensity i.e. fraction of net sown area to total geographical area increase from 36 percent in 1950-51 to 40.5 percent in 1960-61 and 43 percent by 1970-71 where it has since stabilized.

- Cropping intensity i.e. gross sown area as percent of net sown area increased from 111 percent in 1950-51 to 115 percent in 1960-61, 118 percent in 1970—71 and 130 percent by mid 1990s.
- While the contribution of increased area in the growth of agriculture has declined over time, that of productivity has increased. The yield of all crops grew at 1.5 percent per annum between early 1950s and mid 1960s. the pace accelerated to 1.7 percent in the 1970s and then

to 3 percent per annum between early 1980s and mid 90s. Unlike the gains in area, which benefited non-foodgrains, the gains in productivity accrued mostly to foodgrains.

- India's forest resources have been dwindling. According to the 'State of Forest Report' (1997) the total forest cover of the country is estimated at 63.34 million ha i.e. 19.27 percent of the geographic area of the country. Of these the dense forest (crown density more than forty percent) and open forest (crown density 10 to 40 percent) occupying about 11 and 8 percent of the geographic area respectively and mangroves occupy 0.15 percent of the geographic area. The country has lost about 5482 sq. km. of forest cover since the 1995 assessment. By any estimate the area under forest is far below the national policy goals and many areas nominally under forest are being used for non-forest purposes. Similarly 'uncultivated lands' such as permanent pastures, miscellaneous tree crops, cultivable wastes and fallow is subject to increasing competition from uses other than feeding livestock.
- The growth of livestock population is an important source of competition for land. The increase in number of major classes of livestock are shown in *table 13.*
- The area sown to fodder crops is not recorded. Information available from other sources provide an estimate ranging from 4 to 5.5 percent of the net sown area and suggest that the area under fodder crops will have to increase to 10 percent or more to support increasing livestock based activity.

The pressure on India's land and water resources is seriously threatening native plant and animal diversity. India has uniquely rich and diverse genetic base. With increasing agriculture and economic development the genetic pool is declining.

This decline, if unchecked and poorly managed can have unforeseen and adverse consequences for the sustainability of agriculture of the region.

AGRICULTURE IN THE CHANGING GLOBAL SCENARIO

Steady globalization of trade has profound implications for future agricultural development. The diversity of India's agro-ecological setting, high bio-diversity and relatively low cost of labour provide potential for agricultural competitiveness in a globalized economy. It is expected that with increasing globalization of markets over the years there will be demands for agricultural intensification. This will also be favoured because of greater backward and forward linkages between agriculture and food industry.

Therefore, increase in production and productivity are bound to be strategically important to economy. Intensification will not only favour alleviation of rural poverty but will also improve resource conservation particularly in the small farming sector where farmers can be encouraged to take up organized production of high value crops such as fruits, specialty vegetables, flowers medicinal and aromatic herbs etc. Stronger demands for crops of the small farmers' will not only improve incomes and welfare but will also make investments in technology and resource conservation more attractive.

The General Agreement on Tariff and Trade (GATT) and liberalization of global trade is bound to have impact on future land use and production pattern. Understanding the local, national and international environment under which agricultural production is taking shape will be crucial in developing our own strategies.

EXTENSION STRATEGIES

Since early fifties a number of public by funded agricultural development programmes have been sponsored. These have included programmes like the National Extension Service (NES) Blocks in 1953, the Intensive Agricultural District Programme (IADP) in 1961-62, the Intensive Agricultural Area Programme (IAAP) 1964-65, the High Yielding Variety (HYV) programme 1966-67 and the Small and Marginal Farmers' Development Programmes (SMFDP) in 1969-70. Though these programmes had a perceptible impact the efforts did not get replicated over different areas and categories of farmers. In mid seventies based on pilot level project in Rajasthan Canal and Chambal command area a 'Training and Visit' (T&V) system of extension was promoted in different states. Extension efforts of the Indian Council of Agricultural Research through its research Institutes and the State Agricultural University were largely limited to demonstration of new technologies through such programmes as National Demonstration Project, Operational Research Project, the Lab to Land Programme and the Krishi Vigyan Kendras. However, there appears much to be desired in the way that extension programmes are conceived and implemented.

At present extension programmes are implemented in largely a top-down fashion leaving little scope for localized planning and action. Farmers are almost passive receivers and their involvement in the process of technology generation and adoption is almost absent. Extension services, at present, are almost exclusively in the public sector domain and there is no effort or institutional support for other operators e.g. the NGOs, the corporate bodies etc.

Extension programmes sponsored by the government operate largely in isolation and there appears a strong need to view the extension programmes as an integral part of the research and development process.

The challenges facing agricultural development call for fundamental changes in our approach to technology transfer/ extension programmes. Changes are necessary in the context of changing economic environment following policy adjustments in relation to privatization, deregulation and globalization calling for greater efficiency and effectiveness of the extension system. More importantly there is need for

- Greater emphasis on providing producers with knowledge and understanding needed to overcome the problems or to exploit opportunities of their own specific production systems. Correspondingly there will be a need to de-emphasize 'package of practices' or the blanket recommendations, top down approach followed thus far.
- Shift in the focus of public extension systems from promoting inputs use to one on sustainable management

of resources and improvements in the production system as a whole.

- Closer interaction between farmers, extension scientists and production system researchers in diagnosing problems and identifying location specific recommendations emphasizing participation and education rather than being prescriptive.
- Widening the range of extension delivering agencies. While the publicly operated extension systems will continue to be important, there will appear a greater role for NGOs, farmers' associations and corporate sectors in particular situations. Role of commercial suppliers of seeds, agrochemicals, machinery, vaccines and medicines in providing advisories, as is already being done in a limited way, will need to be encouraged and factored into public system's own priorities.
- Wider and more creative use of mass media in tune with current developments in information technology to get information across to the farming community whose ability to overcome constrains at farm level will increasingly depend on access to reliable and up-to-date information.

TECHNOLOGICAL NEEDS AND FUTURE AGRICULTURE

It is apparent that the tasks of meeting the consumption needs of the projected population are going to be more difficult given the higher productivity base than in 1960s. There is also a growing realization that previous strategies of generating and promoting technologies have contributed to serious and widespread problems of environmental and natural resource degradation. This implies that in future the technologies that are developed and promoted must result not only in increased productivity level but also ensure that the quality of natural resource base is preserved and enhanced. In short, they lead to sustainable improvements in agricultural production.

Productivity gains during the 'Green Revolution' era were largely confined to relatively well endowed areas. Given the wide range of agroecological setting and producers, Indian agriculture is faced with a great diversity of needs, opportunities and prospects. Future growth needs to be more rapid, more widely distributed and better targeted. Responding to these challenges will call for more efficient and sustainable use of increasingly scarce land water and germplasm resources.

Technical solutions required to solve problems will be increasingly location-specific and matched to the huge agroecological/climatic diversity. Detailed indigenous knowledge and greater skills in blending modern and traditional technologies to enhance productive efficiency will be more than ever before, key to the farming success and sectoral growth. Most technological solutions will have to be generated and adapted locally to make them compatible with socio-economic conditions of farming community.

New technologies are needed to push the yield frontiers further, utilize inputs more efficiently and diversify to more sustainable and higher value cropping patterns. These are all knowledge intensive technologies that require both a strong research and extension system and skilled farmers but also a reinvigorated interface where the emphasis is on mutual exchange of information bringing advantages to all. At the same time potential of less favoured areas must be better exploited to meet the targets of growth and poverty alleviation.

These challenges have profound implications for products of agricultural research. The way they are transferred to the farmers and indeed the way research is organized and conducted. One thing is, however, clear – the new generation of technologies will have to be much more site specific, based on high quality science and a heightened opportunity for end user participation in the identification of targets. These must be not only aimed at increasing farmers' technical knowledge and understanding of science based agriculture but also taking advantage of opportunities for full integration with indigenous knowledge. It will also need to take on the challenges of incorporating the socio-economic context and role of markets.

With the passage of time and accelerated by macro-economic reforms undertaken in recent years, the Institutional arrangements as well as the mode of functions of bodies responsible for providing technical underpinning to agricultural growth are proving increasingly inadequate. Changes are needed urgently to respond to new demands for agricultural technologies from several directions. Increasing pressure to maintain and enhance the integrity of degrading natural resources, changes in demands and opportunities arising from economic liberalization, unprecedented opportunities arising from advances in biotechnology, information revolution and most importantly the need and urgency to reach the poor and disadvantaged who have been by passed by the green revolution technologies.

Another important implication of increasing globalization relates to the need for greater attention to the quality of produce and products both for the domestic and the foreign markets. This would imply that production must be tuned to actual rapidly changing product demand. Such adaptation to global markets would require state of the art research, which can be achieved only by setting global standards of research, focus on well defined priorities and mechanisms which permit close interaction of farmers with researchers, the private sector and markets.

4

Problems of Indian Agriculture

An important aspect to note about Indian agricultural production (in most parts of the country) is its relationship (even dependency) on skilful and wise water-management practices. One of the unique features of the Indian Subcontinent is how almost its entire rainfall is concentrated in the few monsoon months. During the monsoons, the Indian Subcontinent is usually gifted with bountiful rains, although not infrequently, this bountiful monsoon can turn into a terror, causing uncontrollable floods in parts of the country. Conversely, every few years, the monsoon can be erratic and deficient, leading to drought and the possibility of famine.

Thus, all through history the development of Indian agriculture has been inextricably linked with effective water-management practices that have either been taken up by the state, or by local village communities. Water management has necessitated a certain degree of cooperation and collective spirit in the Indian countryside, and until the imposition of colonial rule, it precluded any widespread development of private property in India.

Regional rulers, or local representatives of the state were generally obliged to allocate a certain percentage of the agricultural taxes on building and managing water-storage, water-harvesting and/or water-diverting structures which facilitated a second crop, and provided water for drinking and other purposes in the long dry season.

Only a small percentage of Indian farmers have enjoyed the luxury of natural irrigation, although there are reports that in certain parts of the country, the soil used to retain enough moisture well beyond the monsoon months. However, it is equally true that the drying up of wells led to mass migrations, and sudden depopulation of old towns and villages.

In Assam, Bengal and Bihar - all flood-prone states, there is evidence of a massive network of canals that allowed both effective drainage to prevent flooding during the heavy monsoon months, and also provide for fishing, transportation and irrigation arteries in the dry seasons.

Intelligent water-management thus allowed for the growth of a healthy agricultural surplus, that in turn facilitated steady urbanization (albeit at a much slower pace than seen in the industrial era), and the development of a variety of pre-industrial manufacturing in areas such as textiles, jewelry, wood/metal-working etc.

Because of the immense importance of effective water-management, the entire revenue system of the state was structured so as to take into account both the necessity of water-management and the inherent dependency that existed between Indian agriculture and the availability of water. Most Indian states attempted to collect revenues in a manner that did not entirely destroy the village solidarity that was essential in proper sharing and management of common water-resources. At the same time, urban tax collectors had to deal with mediating entities from the villages, so that they did not tax at a rate that might lead to the destruction of water-management facilities so essential for life and sustainable agriculture.

By and large, taxes were imposed on villages collectively (not on individual farmers directly), and the village elites (whether Brahmins or others) were obliged to ensure that the burden of taxes did not destroy the complete viability of agriculture. Taxes were also adjusted keeping in mind whether the land was well-irrigated or not.

When ruling dynasties became utterly corrupt and decadent, and failed to tax within limits, or neglected the construction and maintenance of collective water-management facilities, they were confronted with revolts and rebellions, and even risked being overthrown. There is evidence that whenever the tax collectors of a certain regime became too greedy (such as what happened in certain villages/districts during the reigns of some of Northern India's more rapacious invading conquerors), entire villages would be abandoned to escape the burden of heavy taxes. As long as India was not very densely populated and there was the potential of new lands to cultivate, (and the reach of the armed state was not absolute and all-encompassing), this was one of the means by which Indian villagers - both artisans and peasants were able to escape the burden of back-breaking taxes.

But as the population grew and states became more powerful, this escape route was gradually sealed off, and this then led to regional peasant rebellions - most significant of which were the Sikh and the Maratha revolts against the Mughals.

Thus, the British arrived in India at a time when peasant rebellions were beginning to eat away at the invincibility of the might of the Mughal empire. However, even as the decentralization that resulted from the break-up of the widely resented Mughal empire eased slightly the burden on the peasantry in most instances, it also provided an opportunity for the increasingly cunning and shrewd Britishers in India to seize the opportunity, and exploit the widening cleavages in Indian society for their own nefarious ends.

COLONIAL ASPECTS OF EXPLOITATION IN THE INDIAN COUNTRYSIDE

The imposition of Colonial Rule in India led to a drastic break with the past, in that not only did the scale and intensity of the exploitation of the village communities greatly increase, it also led to the introduction of new, and almost entirely parasitic intermediaries between the state and the tax paying masses - intermediaries who were typically induced (or pressurized) to abandon traditional restraint, and discard the old formulae that helped mediate the burden on the typical village peasant or artisan. (It is notable that whereas ancient texts such as the Manusmriti put a cap on agricultural taxes at one-sixth of the produce, and taxes during Mughal rule had risen to something between a third and almost half the produce, during British rule, the peasant was often left with barely a third of the crop.)

Not only were these intermediaries under no pressure to build or maintain traditional water-management structures, the manner in which they were chosen by the British to collect the taxes actually discouraged any diversion of the taxes towards any capital upkeep. Over time, this not only led to the destruction of century-old water management structures, but also to the virtual destruction of the knowledge systems and cultural traditions that had helped build and preserve these water-management techniques over the centuries in states such as Bihar, Bengal, Karnataka, Tamil Nadu and others . (Note that this has been highlighted not only by Rajni-Palme Dutt in *India Today*, but also by several other authors such as Margaret Bourke-White in the *'Famine in South India'* chapter in her book: *"Halfway to Freedom"*).

In addition, the British obliged these intermediaries to collect the taxes **not in kind**, but in **cash**. Since the peasant had little experience or understanding of a cash economy (or any control on the price of grain and agro-inputs), this put a further burden on the peasant, who now also had to face the thuggery of unscrupulous grain traders and usurious money-lenders - who took full advantage of the highly diminished status of the Indian peasant in the colonial dispensation.

(Note that the existence of money-lenders and traders is documented from at least Ashokan times, but the traditional Indian state typically insisted on a cap on interest payments and profits. Moreover, money-lenders and traders were traditionally active mostly in towns, not in villages. The encroachment of the cash economy and the growth of usury in the countryside began to adversely affect the peasants during the later Mughal period, but the problem grew very dramatically during colonial rule.)

As a result, not only was the typical Indian peasant reduced to a state of utter degradation, social relations within the village also became highly distorted. The traditional solidarity that had existed between villagers was now subject to the divisive and ruinous tactics of the parasitic intermediaries who had the protection and support of the colonial state.

As colonial rule progressed, the typical Indian peasant (and village artisan) faced a dual burden. Not only did the traditional feudal form of exploitation greatly intensify (in comparison to traditional Indian practices), the typical Indian villager was now also subject to the merciless forces of mercantile capitalism, who attempted to extract every ounce of savings or assets that any villager might still possess. Eventually, the typical Indian villager was stripped of all savings, and driven to mortgaging a considerable portion of any assets - whether personal jewelry, land and livestock, or tools and equipment. But even as the Indian money-lenders and grain traders fleeced the village artisans and peasants, their accumulated wealth was not especially significant in comparison to the British mercantile capitalists of the East Indian Company who kept a lion's share of the loot.

It is especially important to note that whereas a share of the village surplus went to a few Indian intermediaries (who served as a buffer between the British colonial over-lords and the Indian masses, and generally acted to quash the class struggle), it was the British who expropriated the bulk of the peasant's assets and annual agro-produce, and transferred most of this looted wealth to the British homeland where it was turned into industrial capital and urban development.

This point is of special significance in understanding the unique nature of class contradictions in the Indian countryside (as compared to elsewhere in the non-colonized world.) For instance, almost throughout Europe, the peasantry was also subject to exploitation by the feudal landlords, but the expropriated surplus from the countryside eventually found its way into the urban banking system, and ultimately facilitated in the growth of industrial capitalism. Sections of the impoverished peasantry then had the chance to gradually move into industrial occupations.

But India was subject to an especially tragic plight. Its masses were thoroughly impoverished, but their immediate exploiters were able to keep only a very small share of what had been extracted from the peasantry. This meant that even when a transformation into industrial capitalism became possible (after freedom from British colonial rule) - there was simply not enough accumulated savings to fund the rapid expansion of industry that could eventually transform the countryside.

When the British left, they left India with extremely limited material means to industrialize, but a sizeable class of parasitic intermediaries who would serve as a constant drag on future development. Unlike India's pre-colonial mercantile classes and the more nationalist sections of India's traditional feudal nobles and intermediaries (who were mostly neutralized or exterminated in 1857) this new colonial era mercantile class was extremely backward in cultural terms and was generally lacking in nationalist spirit.

In comparison, the mercantile classes in Europe of the late 18th and 19th centuries were not only wealthier, but also more urbane, and displayed a much higher level of cultural, scientific and technological openness and curiosity. This was also generally true of Japan, and to a somewhat lesser degree, of mercantile classes in other parts of Asia that escaped direct European colonial rule (such as in Korea, Thailand and even China).

It should also be noted that whereas the mercantile classes in Europe, Japan, Korea and China were moving in a broadly secular direction, the Indian mercantile classes were excessively weighed down by sentiments favoring religious revivalism or extreme sectarianism (a trend that was especially strong amongst Muslims, but also posed a serious problem amongst Sikhs and Hindus).

It is thus important to understand that not only was the Indian countryside weighed down by a new and more intensified feudal oppression in the colonial era (such as through the Zamindari or Mahalwari systems), it was also oppressed by a highly reactionary and socially conservative intermediary class of petty mercantile capitalists who would pose as the biggest obstacle to any radical or revolutionary developments in the Indian countryside.

Finally, it should be noted that colonial rule exacerbated the problems of the Indian countryside in yet another way. Prior to colonization, India was steadily becoming more urbanized, with some estimates suggesting that as much as a third of the Indian population may have been living in large or small towns. In addition, even in the villages, a considerable proportion of the population comprised of skilled artisans - weavers, potters, carpenters, metal-workers, painters etc. The proportion of the population that was exclusively engaged in agricultural production was steadily decreasing.

But in the colonial regime, several laws were passed that led to a catastrophic de-urbanization and de-industrialization of India. Trade tariffs and excise duties were set so as to destroy Indian industries, and squeeze domestic trade. In states like Bihar and Bengal, severe restrictions were placed on the use of inland water-ways - causing fishing and inland shipping and transportation to suffer. This led to even greater pressures on agriculture since large categories of highly skilled artisans and non-agricultural workers were thrown out of work.

EFFECTS OF PARTITION

The problem for India was compounded further by the demands for partition, and the insistence of the Muslim League in splitting the country on communal lines. It is important to note that prior to partition, the best naturally irrigated and most productive lands in Northern India were in North Western Punjab (now Pakistan). Fed by a network of all-weather rivers, no other part of the Indian sub-continent was capable of producing more grain with relatively little capital investment. Similarly, Bangladesh was a much better irrigated land than West Bengal (although it faced far greater risks of flooding than Pakistan Punjab).

But in any case, it is worth pointing out that the bulk of pre-partition India's wheat, cotton and jute were grown in what are now Pakistan and Bangladesh. Not only did partition provide India with a smaller share of the land (relative to the ratios of the Hindu, Sikh and Muslim populations), it also left India without the means to feed itself, and without some of the essential raw materials for industry. (Although India did inherit bulk of the cotton and jute mills, India was unable to take full advantage of this until it had increased its own production of jute and cotton to supply these mills, whereas the raw material producers in West and East Pakistan could at least export their raw materials)

It should also be noted that partition disrupted what had been a somewhat natural distribution of labor between what became Pakistan and India. Since West Punjab and Sindh were better suited to agriculture, the relatively arid states of Rajasthan and Gujarat specialized in all manner of cottage industries and manufacturing that needed the raw materials and markets of West Punjab and Sindh to flourish.

In the same way, there was considerable co-dependencies not only between Western and Eastern Bengal, but also between undivided Bengal and its neighboring states such as Assam, Orissa and Bihar. The artificial vivisection of Bengal virtually destroyed the potential of regenerating the traditional trade linkages that had been so vital to the prosperity of the Eastern half of India. Geographically, it should be evident that the management of water and other natural resources cannot be efficiently undertaken without the effective coordination and active involvement of all the states in the Ganga-Brahmaputra region.

But the vitiated atmosphere under which partition took place meant that the possibilities of mutually beneficial cooperation in matters of resource-sharing, trade, transportation and agriculture were practically foreclosed. (To this day, communal forces in Bangladesh have resisted attempts by India to expand cooperation since they would rather see India suffer more rather than see both India and Bangladesh make more rapid progress.)

Thus, when the British left, India had become a much more village-based agricultural economy than it had been earlier - it is likely that the British had pushed India back by several centuries in this respect. Yet, its prospects for rapid industrialization were also quite bleak. India had inherited one of the most depressing scenarios in Asia - a fairly densely populated nation with a pitiful urban base, a rural infrastructure in wrecks, a huge mass of population forced to survive exclusively on agricultural production - yet cynically oppressed by petty mercantile capitalists and feudal intermediaries. Moreover, it had also been left artificially divided on communal lines that meant that important historical linkages might never be revived (or revived only with great difficulty).

No civilized nation could have faced a worse set of circumstances.

THE INDIAN POLITICAL SITUATION POST-1947

The Indian National Congress conducted a dual (and somewhat contradictory) policy relative to Indian agriculture. On the one hand, it attempted to ameliorate the problems of the Indian countryside by investing in irrigation, agro-research, agro-modernization and mechanization, but on the other hand, it was only half-hearted (at best) in implementing its own platform of land reforms because it was reluctant to hurt the interests of the feudal and mercantile intermediaries that oppressed most peasants.

Where the struggle for 'land to the tiller' became very intense, it was obliged to make concessions to the impoverished peasants, and move against the parasitic forces, but wherever the demands for land reform were not backed up by strong and effective people's movements, it chose to drag its feet.

It is only in the sixties, when India's Naxal movement began to spread and acquire a certain degree of political depth did the Indira Congress feel the threat and necessity of clipping the wings of the Zamindars, taking away the privy purses of the Maharajas, and assisting farmers more actively. Even as the Naxal movement was harshly repressed, banks were nationalized and were required to set up more branches in rural areas so that loans for seeds and farm equipment (and inputs such as fertilizers and pesticides) could be purchased through loans at more reasonable interest rates. The state also actively intervened in the buying and selling of grain.

Although these measures did not help all sections of the peasantry, it did allow a section of the country's farmers to gradually liberate themselves from the clutches of the old parasitic intermediaries, and move towards mechanized farming. Over time, peasants struggles put further pressure on the government leading to periodic waiving of loans, and easing of credit norms so that even middle farmers could get loans from government banks and rural credit unions. Subsidies were also provided for electricity and water use, as well as on the sale price of fertilizers and pesticides.

Thus, a combination of people's pressure and the gradual expansion of industrial capitalism has brought about significant changes in the Indian countryside. Whereas a section of Indian farmers (at least in some states) have become rich capitalist farmers, a large middle layer of farmers has also developed which is able to survive (even eke out a small surplus) in the good agricultural years (when the monsoon and market conditions happen to be favorable). These middle-income farmers also hire seasonal labor - and this further distinguishes them from the poorest farmers who can't even survive in the good years.

Thus today, there is a growing differentiation in the Indian countryside, and this is not too different from what has been seen in many other relatively mature capitalist countries. However, the increasing sub-division of land, and the rising costs of agricultural inputs is a constant source of problems and tensions in the countryside.

Comparisons with Europe

In many respects, developments in Indian agriculture have paralleled developments in the Germany and Italy in an earlier century. In both Germany and Italy, the transformation towards a modern industrial society took place slowly, and through a series of evolutionary steps - as is occurring in India today.

The main differences between India and Germany and Italy are simply that not only did Germany and Italy escape the horrors of colonization, they indirectly benefited from the colonial loot. This meant that as their urban industrial economies expanded, these nations were able to absorb their landless populations in their rapidly expanding industrial and services sectors. When industrial expansion wasn't fast enough, Europe's landless had another escape route - external migration. For more than a century, Europe's unemployed and poor escaped to the US, Canada, South America, South Africa, Australia and New Zealand - where the indigenous populations were killed (or displaced from the best agricultural lands).

Whereas Europe had the distinct advantage of being able to export away its rural poor to land it grabbed from indigenous communities (who were not well-armed or skilled in modern warfare), India cannot simply wish away its excess rural population.

SOLUTIONS TO THE PROBLEMS OF INDIAN AGRICULTURE

The subdivision of already small farms in an agricultural model whose profitability is greatly linked to economies of scale makes the survival of the small farmer very difficult. Although the introduction of tractors and other industrial implements, of chemical fertilizers and pesticides, greatly increased the productivity of land and labor, it has not come without considerable burdens for the small farmer who is finding it very hard to compete with large and medium farmers. In most industrial nations, the introduction of such practices in the countryside has been accompanied by a gradual reduction in the rural population so that those who remain on the land can enjoy so

Art, science, and industry of managing the growth of plants and animals for human use. In a broad sense agriculture includes cultivation of the soil, growing and harvesting crops, breeding and raising of livestock, dairying, and forestry (see Animal Husbandry ; Crop Farming ; Dairy Farming ; Forestry ; Poultry Farming ; Soil Management).

Regional and national agriculture are covered in more detail in individual continent and country articles. See also separate articles on the states of Australia and the U.S. and the provinces of Canada.

Modern agriculture depends heavily on engineering and technology and on the biological and physical sciences. Irrigation, drainage, conservation (qq.v.), and sanitation (see Engineering: Sanitary Engineering) each of which is important in successful farming-are some of the fields requiring the specialized knowledge of agricultural engineers.

Agricultural chemistry deals with other vital farm problems, such as uses of fertilizer (q.v.), insecticide, and fungicide (see Fungicides; Pest Control), soil makeup, analysis of agricultural products, and nutritional needs of farm animals.

Plant breeding (q.v.) and genetics (q.v.) contribute immeasurably to farm productivity. Genetics has also placed livestock breeding on a scientific basis. Hydroponics (q.v.), a method of soiless gardening in which plants are grown in chemical nutrient solutions, may solve additional agricultural problems.

The packing, processing, and marketing of agricultural products are closely related activities also influenced by science. Methods of quick-freezing and dehydration have increased the markets for farm products (see Food Processing and Preservation ; Meat-packing Industry).

Mechanization, the outstanding characteristic of late 19th and 20th-century agriculture, has eased much of the backbreaking toil of the farmer. More significantly, mechanization has enormously increased farm efficiency and productivity. See Agricultural Machinery .

Airplanes and helicopters are employed in agriculture for such purposes as seeding, transporting perishable products, and fighting forest fires, and in spraying operations involved in insect and disease control. Radio and television disseminate vital weather reports and other information that is of concern to farmers.

COMPARISION WITH WORLD AGRICULTURE

Over the 10,000 years since agriculture began to be developed, peoples everywhere have discovered the food value of wild plants and animals and domesticated and bred them. The most important are cereals such as wheat, rice, barley, corn, and rye; sugarcane and sugar beets; meat animals such as sheep, cattle, goats, and pigs or swine; poultry such as chickens, ducks, and turkeys; and such products as milk, cheese, eggs, nuts, and oils. Fruits, vegetables, and olives are also major foods for people; feed grains for animals include soybeans, field corn, and sorghum. Separate articles on individual plants and animals contain further information. See also Grasses ; Hay ; Legume ; Silage .

Agricultural income is also derived from nonfood crops such as rubber, fiber plants, tobacco, and oilseeds used in synthetic chemical compounds, as well as raising animals for pelt.

The conditions that determine what will be raised in an area include climate, water supply, and terrain. See also Climate; Ecology; Water Supply and Waterworks .

Nearly 50 percent of the world's labor force is employed in agriculture. The distribution in the late 1980s ranged from 64 percent of the economically active population in Africa to less than 4 percent in the U.S. and Canada. In Asia the figure was 61 percent; in South America, 24 percent; in Eastern Europe and the Soviet Union, 15 percent, and in Western Europe, 7 percent.

Farm size varies widely from region to region. In the late 1980s, the average for Canadian farms was about 230 ha (about 570 acres) per farm; for U.S. farms, about 185 ha (about 460 acres). The average size of a single landholding in the Philippines, however, may be somewhat less than 3.6 ha (less than 9 acres), and in Indonesia, a little less than 1.2 ha (less than 3 acres).

Size also depends on the purpose of the farm. Commercial farming, or production for cash, is usually on large holdings. The latifundia of Latin America are large, privately owned estates worked by tenant labor. Single-crop plantations produce tea, rubber, and cocoa. Wheat farms are most efficient when they comprise some thousands of hectares and can be worked by teams of people and machines. Australian sheep stations and other livestock farms must be large to provide grazing for thousands of animals. The agricultural plots of Chinese communes and the cooperative farms held by Peruvian communities are other necessarily large agricultural units, as were the collective farms that were owned and operated by state employees in the former Soviet Union.

Individual subsistence farms or small-family mixed-farm operations are decreasing in number in developed countries but are still numerous in the developing countries of Africa and Asia. A "back-to-the-land" movement in the U.S. reversed the decline of small farms in New England and Alaska in the decade from 1970 to 1980.

Nomadic herders range over large areas in sub-Saharan Africa, Afghanistan, and Lapland; and herding is a major part of agriculture in such areas as Mongolia.

Much of the foreign exchange earned by a country may be derived from a single commodity; for example, Sri Lanka depends on tea, Denmark specializes in dairy products, Australia in wool, and New Zealand and Argentina in meat products. In the U.S., wheat, corn, and soybeans have become major foreign exchange commodities in recent decades.

The importance of an individual country as an exporter of agricultural products depends on many variables. Among them is the possibility that the country is too little developed industrially to produce manufactured goods in sufficient quantity or technical sophistication. Such agricultural exporters include Ghana, with cocoa, and Burma (Myanmar), with rice. On the other hand, an exceptionally well-developed country may produce surpluses that are not needed by its own population; such has been the case of the U.S., Canada, and some of the Western European countries.

Because nations depend on agriculture not only for food but for national income and raw materials for industry as well, trade in agriculture is a constant international concern. It is regulated by international agreements such as the General Agreement on Tariffs and Trade and by trading areas such as the European Community.

The Food and Agricultural Organization (FAO) of the UN directs much attention to agricultural trade and policies. According to the FAO, world agricultural production, stimulated by improving technology, reached a record high in the late 1980s. Further, agricultural output in developing nations increased 41 percent during the 1977-88 period, as compared to a rise of 9 percent in developed countries. On a per capita basis, however, food production rose by only 12 percent in developing nations, and less than 1 percent in developed countries.

Early agriculturists were, it is agreed, largely of Neolithic culture. Sites occupied by such people are located in southwestern Asia, in what are now Iran, Iraq, Israel, Jordan, Syria, and Turkey; in southeastern Asia, in what is now Thailand; in Africa, along the Nile River in Egypt; and in Europe, along the Danube River and in Macedonia, Thrace, and Thessaly. Early centers of agriculture have also been identified in the Huang He (Yellow River) area of China; the Indus River valley of India and Pakistan; and the Tehuacan Valley of Mexico, northwest of the Isthmus of Tehuantepec.

The dates of domesticated plants and animals vary with the regions, but most predate the 6th millennium bc, and the earliest may date from 10,000 bc. Scientists have carried out carbon-14 testing of animal and plant remains and have dated finds of domesticated sheep at 9000 bc in northern Iraq; cattle in the 6th millennium bc in northeastern Iran; goats at 8000 bc in central Iran; pigs at 8000 bc in Thailand and 7000 bc in Thessaly; onagers, or asses, at 7000 bc in Jarmo, Iraq; and horses at 4350 bc in Ukraine. The llama and alpaca were domesticated in the Andean regions of South America by the middle of the 3d millennium bc .

According to carbon dating, wheat and barley were domesticated in the Middle East in the 8th millennium bc; millet and rice in China and southeastern Asia by 5500 bc; and squash in Mexico about 8000 bc. Legumes found in Thessaly and Macedonia are dated as early as 6000 bc. Flax was grown and apparently woven into textiles early in the Neolithic period.

The farmer began, most probably, by noting which of the wild plants were edible or otherwise useful and learned to save the seed and to replant it in cleared land. Long cultivation of the most prolific and hardiest plants yielded a stable strain.

Herds of goats and sheep were assembled from captured young wild animals, and those with the most useful traits-such as small horns and high milk yield-were bred. The aurochs seems to have been the ancestor of European cattle, and an Asian wild ox of the zebu, the humped cattle of Asia. The cat, dog, and chicken were domesticated very early. The transition from hunting and food gathering to a dependence on food production was gradual, and in a few isolated parts of the world has not yet been accomplished. Crops and domestic meat supplies were augmented by fish and wildfowl as well as by the meat of wild animals.

The Neolithic farmers lived in simple dwellings-in caves and in small houses of sun-baked mud brick or of reed and wood. These homes were grouped into small villages or existed as single farmsteads surrounded by fields, sheltering animals and humans in adjacent or joined buildings. In the Neolithic period, the growth of cities such as Jericho (founded c. 9000 bc) was stimulated by the production of surplus crops.

Pastoralism may have been a later development. Evidence indicates that mixed farming, combining cultivation of crops and stock raising, was the most common Neolithic pattern. Nomadic herders, however, roamed the steppes of Europe and Asia, where the horse and camel were domesticated.

The earliest tools of the farmer were made of wood and stone. They included the stone adz; the sickle or reaping knife with sharpened stone blades, used to gather grain; the digging stick, used to plant seeds, and, with later adaptations, as a spade or hoe; and a rudimentary plow, a modified tree branch used to scratch the surface of the soil and prepare it for planting. The plow was later adapted for pulling by oxen.

The hilly areas of southwestern Asia and the forests of Europe had enough rain to sustain agriculture, but Egypt depended on the annual floods of the Nile to replenish soil moisture and fertility. The inhabitants of the so-called Fertile Crescent, around the Tigris and Euphrates rivers, also depended on annual floods to supply irrigation water. Drainage was necessary to prevent the carrying off of land from the hillsides through which the rivers ran. The farmers who lived in the area near the Huang He developed a system of irrigation and drainage to control the damage caused to their fields in the floodplain of the meandering river.

Although the Neolithic settlements were more permanent than the camps of hunting populations, villages had to be moved periodically in some areas, as the fields lost their fertility from continuous cropping. This was most necessary in northern Europe, where fields were produced by the slash-and-burn method of clearing. The settlements along the Nile, however, were more permanent, because the river deposited fertile silt annually.

Historical Agriculture Through the Roman Period

With the close of the Neolithic period and the introduction of metals, the age of innovation in agriculture was largely over. The historical period-known through written and pictured materials, including the Bible, Near Eastern records and monuments, and Chinese, Greek, and Roman writings-was devoted to improvement. A few high points must serve to outline the development of worldwide agriculture in this era, roughly defined as 2500 bc to ad 500. For a similar period of development in Central and South America, somewhat later in date, see American Indians .

Some plants became newly prominent. Grapes and wine were mentioned in Egyptian records about 2900 bc, and trade in olive oil and wine was widespread in the Mediterranean area in the 1st millennium bc. Rye and oats were cultivated in northern Europe about 1000 bc.

Many vegetables and fruits, including onions, melons, and cucumbers, were grown by the 3d millennium bc in Ur. Dates and figs were an important source of sugar in the Near East, and apples, pomegranates, peaches, and mulberries were grown in the Mediterranean area. Cotton was grown and spun in India about 2000 bc, and linen and silk were used extensively in 2d-millennium China. Felt was made from the wool of sheep in Central Asia and the Russian steppes.

The horse, introduced to Egypt about 1600 bc, was already known in Mesopotamia and Asia Minor. The ox-drawn four-wheeled cart for farm work and two-wheeled chariots drawn by horses were familiar in northern India in the 2nd millennium bc.

Improvements in tools and implements were particularly important. Metal tools were longer lasting and more efficient, and cultivation was greatly improved by such aids as the ox-drawn plow fitted with an iron-tipped point, noted in the 10th century bc in Palestine. In Mesopotamia in the 3d millennium bc a funnel-like device was attached to the plow to aid in seeding, and other early forms of drills were used in China. Threshing was done with animal power in Palestine and Mesopotamia, although reaping, binding, and winnowing were still done by hand. Egypt retained hand seeding through this period, on individual farm plots and large estates alike.

Storage methods for oil and grain were improved. Granaries-jars, dry cisterns, silos, and bins of one sort or another containing stored grain-supported city populations. Indeed, without adequate food supplies and trade in food and nonfood items, the high civilizations of Mesopotamia, northern India, Egypt, and Rome would not have been possible.

Irrigation systems in China, Egypt, and the Near East were elaborated, putting more land into cultivation. The forced labor of peasants and the bureaucracy built up to plan and supervise the work of irrigation were probably basic in the development of the city-states of Sumer. Windmills and water mills, developed toward the end of the Roman period, increased control over the many uncertainties of weather. The introduction of fertilizer, mostly animal manures, and the rotation of fallow and crop land made agriculture more productive.

Mixed farming and stock raising were flourishing in the British Isles and on the continent of Europe as far north as Scandinavia at the beginning of the historical period, already displaying a pattern that persisted throughout the next 3000 years. According to region, fishing and hunting supplemented the food grown by agriculturists.

Shortly after the time of Julius Caesar, the Roman historian Cornelius Tacitus described the "Germans" as a tribal society of free peasant warriors, who cultivated their own lands or left them to fight. About 500 years later, a characteristic European village had a cluster of houses in the middle, surrounded by rudely cultivated fields comprising individually owned farmlands; and meadows, woods, and wasteland were used by the entire community. Oxen and plow were passed from one field to another, and harvesting was a cooperative effort.

Rome appears to have started as a rural agricultural society of independent farmers. In the 1st millennium bc, after the city was established, however, agriculture started a capitalistic development that reached a peak in the Christian era. The large estates that supplied grain to the cities of the empire were owned by absentee landowners and were cultivated by slave labor under the supervision of hired overseers. As slaves, usually war captives, decreased in number, tenants replaced them. The late Roman villa of the Christian era approached the medieval manor in organization; slaves and dependent tenants were forced to work on a fixed schedule, and tenants paid a predetermined share to the estate owner. By the 4th century ad , serfdom (q.v.) was well established, and the former tenant was attached to the land.

Feudal Agriculture

The feudal period in Europe began soon after the fall of the Roman Empire, reaching its height about ad 1100. This period was also that of the development of the Byzantine Empire and of the power of the Saracens in the Middle East and southern Europe. Spain, Italy, and southern France, in particular, were affected by events outside continental Europe.

In the Arab period in Egypt and Spain, irrigation was extended to previously sterile or unproductive land. In Egypt, grain production was sufficient to allow the country to sell wheat in the international market. In Spain, vineyards were planted on sloping land, and irrigation water was brought from the mountains to the plains. In some Islamic areas, oranges, lemons, peaches, and apricots were cultivated.

Rice, sugarcane, cotton, and such vegetables as spinach and artichokes, as well as the characteristic Spanish flavoring saffron, were produced. The silkworm was raised, and its food, the mulberry tree, was grown.

By the 12th century agriculture in the Middle East was static, and Mesopotamia, for example, fell back to subsistence level when its irrigation systems were destroyed by the Mongols. The Crusades increased European contact with Islamic lands and familiarized western Europe with citrus fruits and silk and cotton textiles.

The structure of agriculture was not uniform. In Scandinavia and eastern Germany, the small farms and villages of previous years remained. In mountainous areas and in the marshlands of Slavic Europe, the manorial system could not flourish. Stock raising and olive and grape culture were normally outside the system.

A manor required roughly 350 to 800 ha (about 900 to 2000 acres) of arable land and the same amount of other prescribed lands, such as wetlands, woodlots, and pasture. Typically, the manor was a self-contained community. On it was the large home of the holder of the fief-a military or church vassal of rank, sometimes given the title lord-or of his steward. A parish church was frequently included, and the manor might make up the entire parish. One or more villages might be located on the manor, and village peasants were the actual farmers. Under the direction of an overseer, they produced the crops, raised the meat and draft animals, and paid taxes in services, either forced labor on the lord's lands and other properties or forced military service.

A large manor had a mill for grinding grain, an oven for baking bread, fishponds, orchards, perhaps a winepress or oil press, and herb and vegetable gardens. Bees were kept to produce honey.

Woolen garments were produced from sheep raised on the manor. The wool was spun into yarn, woven into cloth, and then sewn into clothing. Linen textiles could also be produced from flax, which was grown for its oil and fiber.

The food served in a feudal castle or manor house varied-depending on the season and the lord's hunting prowess.

Hunting for meat was, indeed, the major nonmilitary work of the lord and his military retainers. The castle residents could also eat domestic ducks, pheasants, pigeons, geese, hens, and partridges; fish, pork, beef, and mutton; and cabbages, turnips, carrots, onions, beans, and peas. Bread, cheese and butter, ale and wine, and apples and pears also appeared on the table. In the south, olives and olive oil might be used, often instead of butter.

Leather was produced from the manor's cattle. Horses and oxen were the beasts of burden; as heavier horses were bred and a new kind of harness was developed, they became more important. A blacksmith, wheelwright, and carpenter made and maintained crude agricultural tools.

The cultivation regime was rigidly prescribed. The arable land was divided into three fields: one sown in the autumn in wheat or rye; a second sown in the spring in barley, rye, oats, beans, or peas; and the third left fallow. The fields were laid out in strips distributed over the three fields, and without hedges or fences to separate one strip from another. Each male peasant head of household was allotted about 30 strips. Helped by his family and a yoke of oxen, he worked under the direction of the lord's officials. When he worked on his own fields, if he had any, he followed village custom that was probably as rigid as the rule of an overseer.

About the 8th century a 4-year cycle of rotation of fallow appeared. The annual plowing routine on 400 ha would be 100 ha plowed in the autumn and 100 in the spring, and 200 ha of fallow plowed in June. These three periods of plowing, over the year, could produce two crops on 200 ha, depending on the weather. Typically, ten or more oxen were hitched to the tongue of the plow, often little more than a forked tree trunk. The oxen were no larger than modern heifers. At harvest time, all the peasants, including women and children, were expected to work in the fields. After the harvest, the community's animals were let loose on the fields to forage.

Some manors used a strip system. Each strip, with an area of roughly 0.4 ha (about 1 acre), measured about 200 m (about 220 yd) in length and from 1.2 to 5 m (4 to 16.5 ft) in width.

The lord's strips were similar to those of the peasants distributed throughout good and bad field areas. The parish priest might have lands separate from the community fields or strips that he worked himself or that were worked by the peasants.

In all systems, the lord's fields and needs came first, but about three days a week might be left for work on the family strips and garden plots. Wood and peat for fuel were gathered from the commonly held woodlots, and animals were pastured on village meadows. When surpluses of grain, hides, and wool were produced, they were sent to market.

About 1300 a tendency to enclose the common lands and to raise sheep for their wool alone first became apparent. The rise of the textile industry made sheep raising more profitable in England, Flanders, Champagne, Tuscany, Lombardy, and the region of Augsburg in Germany. At the same time, regions about the medieval towns began to specialize in garden produce and dairy products. Independent manorialism was also affected by the wars of 14th- and 15th-century Europe and by the widespread plague outbreaks of the 14th century. Villages were wiped out, and much arable land was abandoned. The remaining peasants were discontented and attempted to improve their conditions.

With the decline in the labor force, only the best land was kept in cultivation, and in southern Italy, for instance, irrigation helped to increase production on the more fertile soils. The emphasis on grain was replaced by diversification, and items requiring more care were produced, such as wine, oil, cheese, butter, and vegetables.

Scientific Agriculture

By the 16th century, population was increasing in Europe, and agricultural production was again expanding.

The nature of agriculture there and in other areas was to change considerably in succeeding centuries. Several reasons can be identified. Europe was cut off from Asia and the Middle East by an extension of Turkish power. New economic theories were being put into practice, directly affecting agriculture. Also, continued wars between England and France, within each of these countries, and in Germany consumed capital and human resources.

A new period of exploration and colonization was undertaken to circumvent Turkey's control of the spice trade, to provide homes for religious refugees, and to provide wealth for European nations convinced that only precious metals constituted wealth.

Colonial agriculture was carried out not only to feed the colonists but also to produce cash crops and to supply food for the home country. This meant cultivation of such crops as sugar, cotton, tobacco, and tea and production of animal products such as wool and hides. From the 15th to the 19th century the slave trade provided needed laborers, replacing natives killed by unaccustomed hard labor in unfavorable climates and substituting for imported Europeans on colonial plantations that required a larger labor force than the colony could provide. Slaves from Africa worked, for instance, in the Caribbean area on sugar plantations and in North America on indigo and cotton plantations. Indians were virtually enslaved in Mexico. Indentured slaves from Europe, and especially from the prisons of England, provided both skilled and unskilled labor to many colonies. Ultimately, however, both slavery and serfdom were substantially wiped out in the 19th century. See Peonage; Plantation; Slavery .

When encountered by the Spanish conquistadors, the more advanced Indians in the New World had intensive agricultural economies but no draft or riding animals and no wheeled vehicles. Squash, beans, peas, and corn had long since been domesticated. Land was owned by clans and other kinship groups or by ruling tribes that had formed sophisticated governments, but not by individuals or individual families. Several civilizations had risen and fallen in Central and South America by the 16th century. Those met by the Spanish were the Aztec, Inca, and Maya.

The scientific revolution resulting from the Renaissance and the Age of Enlightenment in Europe encouraged experimentation in agriculture as well as in other fields. Trial-and-error efforts in plant breeding produced improved crops, and a few new strains of cattle and sheep were developed.

Notable was the Guernsey cow breed, still a heavy milk producer today. Enclosure was greatly speeded up in the 18th century, and individual landowners could determine the disposition of land and of pasture, previously subject to common use. Crop rotation, involving alternation of legumes with grain, was more readily practiced outside the village strip system inherited from the manorial period. In England, where scientific farming was most efficient, enclosure brought about a fundamental reorganization of landownership. From 1660 on, the large landowners had begun to add to their properties, frequently at the expense of small independent farmers. By Victorian times, the agricultural pattern was based on the relationship between the landowner, dependent on rents; the farmer, producer of crops; and the landless laborer, the "hired hand" of American farming lore. Drainage brought more land into cultivation, and, with the Industrial Revolution, farm machinery was introduced.

It is not possible to fix a clear decade or series of events as the start of the agricultural revolution through technology. Among the important advances were the purposeful selective breeding of livestock, begun in the early 1700s, and the spreading of limestone on farm soils in the late 1700s. Mechanical improvements of the traditional wooden plow began in the mid-1600s with small iron points fastened onto the wood with strips of leather. In 1797, Charles Newbold (1764-1835), a blacksmith in Burlington, N.J., introduced the cast-iron moldboard plow. John Deere, another American blacksmith, further improved the plow in the 1830s and manufactured it in steel. Other notable inventions included the seed drill of the English agriculturist Jethro Tull, developed in the early 1700s and progressively improved for more than a century; the reaper of Cyrus McCormick in 1831; and numerous new horse-drawn threshers, cultivators, grain and grass cutters, rakes, and corn shellers. By the late 1800s, steam power was frequently used to replace animal power in drawing plows and in operating threshing machinery.

The demand for food for urban workers and raw materials for industrial plants produced a realignment of world trade. Science and technology developed for industrial purposes were carried over into agriculture, eventually resulting in the agribusinesses of the mid-20th century.

In the 17th and 18th centuries the first systematic attempts were made to study and control pests. Before this time, handpicking and spraying were the usual methods of pest control. In the 19th century, poisons of various types were developed for use in sprays, and biological controls such as predatory insects were also used. Resistant plant varieties were cultivated; this was particularly successful with the European grapevine, in which the grape-bearing stems were grafted onto resistant American rootstocks to defeat the Phylloxera aphid. See Pest Control .

Improvements in transportation affected agriculture. Roads, canals, and rail lines enabled farmers to obtain needed supplies and to market their produce over a wider area. Food could be protected in transport and shipped more economically than before as a result of rail, ship, and refrigeration developments of the late 19th and early 20th centuries. Efficient use of these developments led to increasing specialization and eventual changes in the location of agricultural suppliers. In the last quarter of the 19th century, for example, Australian and North American suppliers displaced European suppliers of grain in the European market. When grain production proved unprofitable for European farmers, or an area became more urbanized, specialization in dairying, cheese making, and other products was emphasized.

The impetus toward more food production in the era following World War II was a result of a new population explosion. A so-called green revolution, involving selective breeding of traditional crops for high yields, new hybrids, and intensive cultivation methods adapted to the climates and cultural conditions of densely populated countries such as India, temporarily stemmed the pressure for more food. A worldwide shortage of petroleum in the mid-1970s, however, reduced the supplies of nitrogen fertilizer helpful to the success of the new varieties. Simultaneously, erratic weather and natural disasters such as drought and floods reduced crop levels throughout the world. Famine seemed to be imminent in the Indian subcontinent and was common in many parts of Africa south of the Sahara. Economic conditions, particularly uncontrolled inflation, threatened the food supplier and the consumer alike. These problems became the determinants of agricultural change and development. See Energy Supply, World; Environment; Food Supply, World .

AGRICULTURE IN THE U.S.

In North America, agriculture had progressed further before the coming of the Europeans than is commonly supposed.

Until the 19th century, agriculture in the U.S. shared the history of European and colonial areas and was dependent on European sources for seed, stocks, livestock, and machinery, such as it was. That dependency, especially the difficulty in procuring suitable implements, made American farmers somewhat more innovative. They were aided by the establishment of societies that lobbied for governmental agencies of agriculture (see Agriculture, Department of); the voluntary cooperation of farmers through associations (see Cooperatives; National Grangem); and the increasing use of various types of power machinery on the farm. Government policies traditionally encouraged the growth of land settlement. The Homestead Act of 1862 and the resettlement plans of the 1930s were the important legislative acts of the 19th and 20th centuries.

In the 20th century steam, gasoline, diesel, and electric power came into wide use. Chemical fertilizers were manufactured in greatly increased quantities, and soil analysis was widely employed to determine the elements needed by a particular soil to maintain or restore its fertility. The loss of soil by erosion (q.v.) was extensively combated by the use of cover crops (quick-growing plants with dense root systems to bind soil), contour plowing (in which the furrow follows the contour of the land and is level, rather than up and down hills that provide channels for runoff water), and strip cropping (sowing strips of dense-rooted plants to serve as water-breaks or windbreaks in fields of plants with loose root systems).

Selective breeding produced improved strains of both farm animals and crop plants. Hybrids of desirable characteristics were developed; especially important for food production was the hybridization of corn in the 1930s. New uses for farm products, by-products, and wastes were discovered. Standards of quality, size, and packing were established for various fruits and vegetables to aid in wholesale marketing. Among the first to be standardized were apples, citrus fruits, celery, berries, and tomatoes. Improvements in storage, processing, and transportation also increased the marketability of farm products. The use of cold-storage warehouses and refrigerated railroad cars was supplemented by the introduction of refrigerated motor trucks, by rapid delivery by airplane, and by the quick-freeze process of preservation, in which farm produce is frozen and packaged the same day that it is picked. Freeze-drying and irradiation have also reached practical application for many perishable foods.

Scientific methods have begun to be applied to pest control, limiting the widespread use of insecticides and fungicides and applying more varied and targeted techniques. New understanding of significant biological control measures and the emphasis on integrated pest management have made possible more effective control of certain kinds of insects.

Chemicals for weed control have become important for a number of crops, in particular cotton and corn. The increasing use of chemicals for the control of insects, diseases, and weeds has brought about additional environmental problems and regulations that make strong demands on the skill of farm operators.

In the 1980s high technology farming, including hybrids for wheat, rice, and other grains, better methods of soil conservation and irrigation, and the growing use of fertilizers has led to the production of more food per capita, not only in the U.S., but in much of the rest of the world. U.S. farmers, however, still have the advantage of superior private and government research facilities to produce and perfect new technologies.

Government Price-Support Policies

One of the recurring problems of American agriculture in the 20th century has been the tendency of farm income to lag behind increases in the costs of production. The problem began in the 1920s, following a period of exceptional prosperity for U.S. farmers. The period 1910-14 was later taken as a standard for the level of farm prices in relation to the general price level and formed the basis for a concept called parity, aimed at maintaining farming as an essential part of the U.S. economy. After the outbreak of World War I the U.S. became the chief source of food for the warring nations of Europe, with U.S. farmers bringing some 16 million additional ha (about 40 million acres) of land under cultivation and investing heavily in new land and equipment. These measures raised production levels until 1920, when the European demand for U.S. farm products suddenly declined, and prices began a continuing downward spiral.

Although attempts were under way to ease the economic difficulties of the farmer, farm income had not begun to recover when the Great Depression of the 1930s intensified them even more. By 1932 the level of farm prices was only about 65 percent of the 1910-14 average. Farmers continued to produce almost as much as before, and even increased their production in an attempt to maintain their income. That succeeded only in lowering farm prices further. By comparison, manufacturers could control their production, thereby maintaining price levels to a certain degree. Although prices for industrial goods declined, they did not drop as severely as farm prices, so that by 1932 farmers were receiving only 58 cents from the sale of their products for every dollar they had to pay for non-farm items.

The federal government, which had done little in the 1920s to help farmers, initiated remedial programs during the administration of President Franklin D. Roosevelt. One approach was to reduce the supply of basic farm commodities. The Agricultural Adjustment Act of 1933 provided payments to farmers in return for agreements to curtail their acreage or their production of wheat, cotton, rice, tobacco, corn, hogs, and dairy products. The act was declared unconstitutional in 1936, but in 1938, after several changes in the membership of the U.S. Supreme Court, a second Agricultural Adjustment Act was passed under which production quotas were set as before.

Payments were financed from taxes imposed on processors and were based on the parity concept.

The government also lent money to farmers to enable them to withhold crops from the market when prices were low and to store the produce so that it might be available in poor crop years.

A third method to limit production provided payments for shifting acreage of soil-depleting crops such as corn, wheat, cotton, tobacco, and rice to soil-conserving plants such as grasses and legumes and for carrying out soil-building practices. In 1939, an all-risk crop insurance program was initiated for interested farmers to prevent economic distress in case of crop failure for hail, floods, and other natural disasters.

Until World War II the problem of low farm prices was not basically a result of overproduction. Rather, it was a consequence of the cycles of business and weather, and of problems of internal distribution, transportation, and credit. Following World War II, however, overproduction became a serious problem. Both during and immediately after the war, farm prices were generally high. Because production costs also were high, parity payments remained in force. Federal transactions in surplus commodities, principally the sale of such commodities at prices less than those paid to farmers, proved costly for the government. To reduce costs of the federal farm program, the administration of President Dwight D. Eisenhower proposed the substitution of flexible or variable price supports for the rigid 90 percent of parity that was in force. A bill authorizing a sliding scale of payments at 82.5 percent to 90 percent of parity on the basic commodities was enacted by the U.S. Congress in 1954.

The Agricultural Act of 1956, otherwise known as the soil-bank program, authorized federal payments to farmers if they reduced production of certain crops. A subsidy plan was formulated whereby farmers would be paid for converting part of their cropland to soil-conserving uses. In practice, the farmers shared the costs of planting trees or grasses and received annual payments compensating them for the economic loss incurred by the removal of some of their land from production.

The Department of Agriculture in the administrations of Presidents John Kennedy and Lyndon Johnson during the 1960s made control of overproduction a primary goal of farm policy. Farmers were offered what was in effect a rental payment for a part of their land that would be taken out of production during the following year. At the same time, measures were undertaken to expand the export market for agricultural products. During this period the ratio of a farmer's per capita income to that of a non-farm person increased from about 50 percent to about 75 percent.

Direct subsidies for withholding agricultural land from production were phased out in 1973, as a result of a proposal by President Richard M. Nixon. In the same year, net farm income swelled to $33.3 billion.

Poor grain harvests throughout the world, particularly in the Soviet Union, prompted massive sales of U.S. government-owned grain reserves. World climatic conditions also helped keep demand for U.S. produce high through the mid-1970s. Toward the end of the decade, exports lessened, prices dropped, and farm income began to fall without a corresponding decrease in costs of production. U.S. net farm income in 1976 fell to $18.7 billion.

In 1978, a limited, voluntary output restriction was begun by President Jimmy Carter. Called the "farmer-held grain reserve program," the action took grains off the market for up to three years or until market prices reached stated levels. The program was intended also to provide an adequate reserve, lessen food-price gyrations and combat inflation, give livestock producers protection from extremes in feed costs, and contribute to greater continuity in foreign food aid.

On Jan. 4, 1980, President Carter declared a limited suspension of grain sales to the Soviet Union in response to that country's invasion of Afghanistan. Additional restrictions included a prohibition on sales of U. S. phosphate. Despite the grain embargo, the U.S. continued to honor a 5-year agreement already in effect that committed it to sell 8 million tons of grain to the Soviets yearly. The year 1980 was an election year, and despite efforts by President Carter's opposition to void the embargo, it continued. Administration officials argued that the Soviets had never been a major customer or even a reliable buyer. U.S. farmers maintained, however, that the action was at their expense and had made 1980 one of their worst years. In fact, U.S. farm exports in 1980 reached an all-time high of $40 billion, but the continued rise in costs of production and an extremely hot summer with accompanying droughts affected many farmers adversely. A new crop insurance program, passed by Congress in the fall of 1980, offered relief from such conditions rather than having to rely on disaster loans, which amounted to $30 million for feed alone in that year.

Whether the 1980 grain embargo had a strong effect on the USSR was a matter of conjecture. Beef production dropped 16 percent, pork was off 10 percent, and milk production fell 4 percent, but by the end of the year the Soviets had apparently obtained their needed grain from other sources. When President Ronald Reagan took office in 1981, he lifted the embargo and extended the agreement that allowed the USSR to purchase 8 million tons of grain yearly from the U.S. The two nations then signed a new 5-year agreement in 1983 that obliged the Soviet Union to import a minimum of 9 million tons of U.S. grain every year.

Farming Regions

The U.S. has ten major farming areas. They vary by soil, slope of land, climate, and distance to market, and in storage and marketing facilities.

The states of the northeast and the Lake states are the country's principal milk-producing areas. Climate and soil there are suited to raising grains and forage for cattle and for pastures. Broiler farming is important to Maine, Delaware, and Maryland. Fruits and vegetables are also important to the region.

The Appalachian region is the major tobacco-producing area of the nation. Peanuts, cattle, and dairy production also are important.

Beef cattle and broilers are the major livestock products farther south in the states of the Southeast; fruit and vegetables and peanuts are also grown. Florida has vast citrus groves and winter vegetable production areas. In the Delta states, principal cash crops are soybeans and cotton. Rice and sugarcane are grown in the more humid and wet areas. With improved pastures, livestock production has gained importance in recent years. It also is a major broiler-producing region.

The Corn Belt, extending from Ohio through Iowa, has rich soil, good climate, and sufficient rainfall for excellent farming. Corn, beef cattle, hogs, and dairy products are of primary importance. Other feed grains, soybeans, and wheat also are grown.

The northern and southern Plains, extending north and south from Canada to Mexico and from the Corn Belt into the Rocky Mountains, are restricted by low rainfall in the western portion and by cold winters and short growing seasons in the north. But about 60 percent of the nation's winter and spring wheat grows in the Plains states. Other small grains, grain sorghums, hay, forage crops, and pastures help make cattle important to the region. Cotton is produced in the southern part.

The Mountain states provide yet a different terrain. Vast areas are suited to cattle and sheep. Wheat is important in the north. Irrigation in the valleys provides water for hay, sugar beets, potatoes, fruits, and vegetables.

The Pacific region includes California, Oregon, and Washington plus Alaska and Hawaii. In the northern mainland, farmers raise wheat, fruit, and potatoes. Dairying, vegetables, and some grain are important to Alaska. Many of the more southerly farmers have large tracts on which they raise vegetables, fruit, and cotton, often under irrigation. Cattle are raised throughout the region. Hawaii grows sugarcane and pineapple as its major crops.

Agricultural Resources

The total land area of the U.S. is about 917 million ha (about 2.27 billion acres), of which about 47 percent is used to produce crops and livestock. The rest is distributed among forestland (29 percent) and urban, transportation, and other uses (24 percent).

Approximately 161 million ha (about 399 million acres) make up cropland resources. Almost 83 percent of cropland is cultivated, including about 23 million ha (about 57 million acres) used for wheat, about 30 million ha (about 74 million acres) used for corn, and about 25 million ha (about 62 million acres) used for hay. More than 50 percent of croplands are prime farmland, the best land for producing food and fiber.

The nation has another nearly 400 million ha (almost 1 billion acres) of nonfederal rural land currently being used for pastures, range, forest, and other purposes. About 27.5 million ha (about 68 million acres) of this land are suitable for conversion to cropland if needed.

Recent Changes

The history of agriculture in the U.S. since the Great Depression has been one of consolidation and increasing efficiency. From a high of 6.8 million farms in 1935, the total number declined to 2.1 million in 1991 on a little less than the same area, about 397 million ha (about 982 million acres). Average farm size in 1935 was about 63 ha (about 155 acres); in 1991 it was about 189 ha (about 467 acres).

About 4.6 million people lived on farms in 1990, based on a new farm definition introduced in 1977 to distinguish between rural residents and people who earned $1000 or more from annual agricultural product sales. The farm population continues to constitute a declining share of the nation's total; about 1 person in every 54, or 1.8 percent, of the nation's 250 million people were farm residents in 1990.

Total value of land and buildings on U.S. farms in 1990 was $658 billion, substantially less than the value in 1980. Value of products sold was $170 billion per year. Overall net farm income was more than $46 billion in 1989, of which government subsidies accounted for 23 percent.

Not including real estate, major expenditures by farmers in 1989 were for feed ($22.7 billion); fuel, lubricants, and maintenance ($13.1 billion); hired labor ($11.9 billion); fertilizer ($7.6 billion); and seed ($3.7 billion). Outstanding farm debt in 1989 was $146 billion, of which about 55 percent was owed on real estate. Interest payments on the mortgage debt were about $7.6 billion per year.

In 1980, a report based on projections by the U.S. government stated that in the next 20 years world food requirements would increase tremendously, with developed countries requiring most of the increase, and food prices would double. Less than five years later, however, the U.S. farmer was enveloped in a major crisis caused by exceptionally heavy farm debts, mounting farm subsidy costs, and rising surpluses. A number of farmers were forced into foreclosure.

The ailing Farm Credit System, a group of 37 farmer-owned banks under the Farm Credit Administration (q.v.) appealed to the government for a $5 to $6 billion fund that would keep the system solvent despite the weak national farm economy. After initial resistance, President Ronald Reagan signed legislation in December 1985 designed to create the Farm Credit System Capital Corp. to take over bad loans from the system's banks and to assume responsibility for foreclosing or restructuring distressed loans.

In December President Reagan also signed the Food Security Act of 1985, legislation designed to govern the nation's farm policies for the next five years, trim farm subsidies, and stimulate farm exports.

5

Indian Agricultural Trade

EXTERNAL TRADE WITH INDIA

Agricultural exports from India were 44 percent of total exports in FY 1960; they decreased to 32 percent in FY 1970, to 31 percent in FY 1980, to 18.5 percent in FY 1988, and to 15.3 percent in FY 1993. This drop in agriculture's share was somewhat misleading because agricultural products, such as cotton and jute, that were exported in the raw form in the 1950s, have been exported as cotton yarn, fabrics, ready-made garments, coir yarn, and jute manufactures since the 1960s.

The composition of agricultural and allied products for export from India changed mainly because of the continuing growth of demand in the domestic market. This demand cut into the surplus available for export despite a continuing desire, on the part of government, to shore up the constant foreign-exchange shortage.

In FY 1960, tea was the principal export by value. Oil cakes, tobacco, cashew kernels, spices, and raw cotton were about equal in value but were only one-eighth of the value of tea exports. By FY 1980, tea was still dominant, but coffee, rice, fish, and fish products came close, followed by oil cakes, cashew kernels, and cotton. In 1992-93 fish and fish products became the primary agricultural export, followed by oil meals, then cereals, and then tea. The share of fish products rose steadily from less than 2 percent of all agricultural exports in FY 1960, to 10 percent in FY 1980, to approximately 15 percent for the three-year period ending in FY 1990, and to 23 percent in FY 1992. The share of tea in agricultural exports fell from 40 percent in FY 1960 to roughly 17 percent in the FY 1988-FY 1990 period, and to only 13 percent by FY 1992.

External Aid for India

Foreign aid—financial and technical—since the 1950s has made a significant contribution to the agricultural progress in rural India. Aid has come from many sources: the United States government, the Ford Foundation, the Rockefeller Foundation, the World Bank, the Food and Agriculture Organization (FAO—see Glossary) of the United Nations (UN), the European Economic Community, the former Soviet Union, Britain, and Japan, among others .

Agricultural aid to India also has come in many forms. Between 1963 and 1972, for example, under a program of the United States Agency for International Development, some 400 American scientists and scholars served on the faculties of India's agricultural universities, while more than 500 faculty members from Indian institutions received advanced training in the United States and other countries. Several hundred agricultural research projects, financed with funds generated from sales of American farm commodities under the United States Public Law 480 program, fueled technological breakthroughs in Indian agriculture.

Aid to the agricultural sector in India continued in the late 1980s and the early 1990s; the FAO, the European Union, the World Bank, and the United Nations Development Programme (UNDP) provided the bulk of the assistance.

The FAO provided technical assistance in a number of emerging areas; it provided quality control for exports; videos for rural communication and training; and market studies for wool processing, mushroom production, and egg and poultry marketing. Operation Flood--a dairy development program—was jointly sponsored by the European Economic Community, the World Bank, and India's National Dairy Development Board (see Livestock and Poultry, this ch.).

The UNDP provided technical assistance by sending foreign experts, consultants, and equipment to India. The World Bank and its affiliates supported agricultural extension, social (community-based) forestry, agricultural credit, dairy development, horticulture, seed development, rain-fed fish farms, storage, marketing, and irrigation.

India has not only been a receiver of aid. Increasingly since independence, India has been sharing its agricultural technology with other developing countries. Numerous foreign scientists have received special and advanced training in India; hundreds of foreign students have attended Indian state agricultural universities. Among other international agricultural endeavors, India has contributed scientists, services, and funds to the work of the International Rice Research Institute, headquartered in the Philippines.

In the late 1980s and early 1990s, India provided short- and long-term training courses to hundreds of foreign specialists each year under a variety of programs, including the Technical Cooperation Scheme of the Colombo Plan for Cooperative Economic and Social Development in Asia and the Pacific and the Technical Cooperation Scheme of the Commonwealth of Nations Assistance Program.

IMPACT OF ECONOMIC REFORMS ON AGRICULTURE IN INDIA

The serious foreign-exchange crisis in 1990 led to a number of well-publicized economic reforms in the early 1990s dealing with trade, industrial licensing, and privatization. The reforms had an impact on the agricultural sector through the central government's effort to withdraw the fertilizer subsidy and place greater emphasis on agricultural exports.

The cut in the fertilizer subsidy was a result of the government's commitment to reduce New Delhi's fiscal deficit by removing grants and subsidies from the budget. The government action led to a reduction in the use of chemical fertilizers and protests by farmers and opposition political parties. The government was forced to continue the subsidies but at a somewhat lower level.

New import and export policies aim at enhancing export capabilities of the agricultural sector by increasing productivity and promoting modernization and competitiveness. The measures to facilitate export growth include allowing the import of capital for the agricultural sector, reducing the list of agricultural products that cannot be exported, and removing the minimum export price from a number of products.

Agricultural exports increased by 30 percent in FY 1991 and 14 percent in FY 1992 in terms of rupee value, but declined by 8 percent from FY 1990 to FY 1992 in United States dollar terms because of the devaluation of the rupee in 1991.

In the mid-1990s, it was expected that agriculture would continue to be the most important sector of the economy for the rest of the decade in terms of the proportion of GDP.

However, even when it is not the sector providing the largest share of GDP, the importance of agriculture is not likely to diminish because of its critical role in providing food, wage goods, employment, and raw materials to industries.

Despite their preoccupation with industrial development, India's planners and policy makers have had to acknowledge the critical role of agriculture in the early 1990s by changing basic policy. The gains in agricultural production should not lead to complacency, however. Continuing increases in productivity, developing allied activities in rural areas, and building infrastructure in rural areas are essential if India is to continue to be self-reliant in food and agricultural products and provide a modest surplus for exports.

AGRICULTURE AND RURAL DEVELOPMENTS

From a nation dependent on food imports to feed its population, India today is not only self—sufficient in grain production, but also has a substantial reserve.

The progress made by agriculture in the last four decades has been one of the biggest success stories of free India. Agriculture and allied activities constitute the single largest contributor to the Gross Domestic Product, almost 33% of it. Agriculture is the means of livelihood of about two—thirds of the work force in the country.

This increase in agricultural production has been brought about by bringing additional area under cultivation, extension of irrigation facilities, the use of improved high yielding variety of seeds, better techniques evolved through agricultural research, water management, and plant protection through judicious use of fertilizers, pesticides and cropping practices.

Crops

The 1970s saw a multi-fold increase in wheat production that heralded the Green Revolution. In the next decade rice production rose significantly; in 1995-96, rice production was 79.6 million tons. Total grain production crossed 191 million tons in 1994-95, a big leap from 51 million tons in 1950-5 1. During the Seventh Plan, the average grain production was 155 million tons, 17 million tons more than the Sixth Plan average.

To carry improved technologies to farmers, a National Pulse Development Program, covering 13 states, was launched in 1986. The Special Food Production Program augmented efforts to boost pulse production further. In 1995-96, pulse production was 13.2 million tons. With some States offering more than the statutory minimum price, sugarcane production also received a boost, in 1995-96 a record 283.0 million tons was registered.

Irrigation

As efforts continued to increase the irrigation potential in the country, the last 40 years saw the gross irrigated area reach 8~ million hectares. Flood forecasting has become an important activity over the years. Over 500 hydrological stations collect and transmit data through 400 wireless stations for issuing forecasts for 157 sites. About 5000 forecasts are issued in a year with 94% accuracy.

The country also receives international support, with the World Bank as a primary source, for developing the water resources. International cooperation is also envisaged in setting up a National Center for Information on Water and Power. As there is a broad seismic belt in the country, particularly along the Himalayan, the Kutch region and ports of Maharashtra, a scheme is being evolved to collect all data on seismic activity at various dam sites.

Fertilizers

The fertilizer industry in India has grown tremendously in the last 30 years. The Government is keen to see that fertilizer reaches the farmers in the remote and hilly areas. It has been decided to decontrol the prices, distribution and movement of phosphatic and potassic fertilizers. Steps have been taken to ensure an increase in the supply of non-chemical fertilizers at reasonable prices.

There are 53 fertilizer quality control laboratories in the country. Since bio—fertilisers are regarded as an effective, cheap and renewable supplement to chemical fertilizers, the Government is implementing a National Project on Development and Use of Bio-fertilisers. Under this scheme, one national and six regional centers for organizing training, demonstrating programs and quality testing of bio-fertilisers has been taken up.

It was a challenging decision of the Government to take Bombay High gas through a 1,700-km pipeline to feed fertilizer plants located in the consumption centers of North India. However, the major policy which has ensured the growth of the fertilizer industry is the thrust on accelerating fertilizer consumption by fixing, on the one hand, low and uniform price for fertilizers, and on the other hand providing the manufacturers adequate compensation through the retention price and subsidy scheme. As expected, fertilizer nutrient demand has gone up from 0.29 million tons in 1960-61 to 13.9 million tons at the end of 1995-96, compared to 12.15MT during 1992-93.

Fisheries

Fish production achieved an all-time high of 4.9 million tons at the end of 1995-96. Programs that have helped boost production include the National Program of Developing Fish Seeds, Fish Farmers' Development Agencies and Brackish Water Fish Farmers' Development Agencies. The Central Institute of Fisheries Nautical and Engineering Training trains the necessary manpower. To diversify fishing methods and introduce processed fish products on a semi-commercial scale, an

Integrated Fisheries Project has been launched. A National Fisheries Advisory Board has also been established.

Food Processing

A Ministry of Food Processing Industries was established in July 1988 to ensure better utilization of farmers' output by inducting modem technology into the processing of food products, thus augmenting the income of farmers and generating employment opportunities in rural areas. A new seeds policy has been adopted to provide access to high quality seeds and plant material for vegetables, fruit, flowers, oil-seeds and pulses, without in any way compromising quarantine conditions. Initiatives have been taken to encourage private sector investment in the food processing industry.

Agricultural Research

The apex body for education, research and extension education in the field of agriculture is the Indian Council of Agricultural Research (ICAR), established in 1929. India's transformation from a food deficit to a food surplus country is largely due to ICAR's smooth and rapid transfer of farm technology from the laboratory to the land.

ICAR discharges its responsibilities through 43 research institutes, four national research bureaus, 20 national research centers, nine project directorates, 70 all-India coordinated research projects, and 109 Krishi Vigyan Kendras (farm science centers). Besides, the program of Agricultural Education is coordinated by ICAR with the curricula and other normative guidance given to the 26 Agricultural Universities and four National Research Institutes.

Oilseeds Production

A Technology Mission on Oilseeds was launched in 1986 to increase production of oilseeds in the country and attain self-sufficiency. Pulses were brought under the Technology Mission in 1990. Before the Mission was launched in 1985-86, oilseed production was 10.83 million tons; during 1995-96, it was estimated at 22.42 million tons, which is a record. Soybean, rapeseed and mustard largely contributed the increase in production. Production of pulses has seen many ups and downs, which is expected to be checked under the Mission. The country grows mainly nine oilseeds, with groundnut, rapeseed and mustard accounting for 62% of total production. Lately, soybean and sunflower have shown major growth potential.

Drinking Water

A Technology Mission on Drinking Water and Related Water Management has been constituted to cover the residual problem villages and provide potable water at 40 liter per capita per day, and 70 liters per capita per day in desert areas inclusive of 30 liters for cattle. The Mission is tackling the problem through 55 mini-missions in project districts and countrywide problem oriented sub-missions. A Village Level Operation and Maintenance (VLOM) pump called India Mark-11 has been developed and is being exported to 40 countries. By March 31, 1993, over 79% of the rural and about 85% of the urban population was provided drinking water facilities.

6

Development in Agriculture and Biotechnologies

SUSTAINABLE SURVIVAL

All living organisms are driven by two motivating forces: to reproduce themselves and to survive. Survival depends largely on protection from their enemies and having access to sufficient food. Consequently, for countless millennia the developmental history of hominids has to do with their search for sustenance and survival: an uneven progression from hunting and gathering to crop cultivation and livestock husbandry, the discovery and devising of biotechnologies by which to increase, protect, preserve and transform natural and cultivated biological species and materials into food, fibre and other goods of personal, domestic or commercial utility.

History records that agriculture has progressed as much through empirical perception as from scientific intention. In today's world, intensive large-scale highly mechanised production systems exist together with simple labour intensive subsistence systems where a few crops and farm animals survive on less than a half-hectare of land. Some nations and regions, which for many centuries were prosperous, maintained efficient profitable agricultural systems and produced well in excess of their needs, subsequently lost or forsook their agricultural capabilities and descended into a state of food insufficiency.

To be sustainable, agriculture requires a complex of critical resources, not least human skills and experience, and the dedicated political will of governments and those who legislate each nation's priorities. Libya and Nigeria are two nations where the discovery of oil has led to gross neglect of agriculture. In other ecologies, changes in weather patterns or excessive exploitation of natural resources and depredation of ecosystems have together degraded once fertile lands into near deserts.

AGRICULTURE AND BIOTECHNOLOGIES

During the first century BCE, a Roman farmer called Varro is reported to have said: "Agriculture is a science which teaches us what crops should be planted in each kind of soil, and what operations are necessary in order that the land will produce the highest yields in perpetuity". The fourth century Bishop Ambrose of Milan wrote in *De Officiis* that agriculture is the only way of making money that gives no offence.

'Agriculture' is derived from the Greek word 'áãñïæ' which means 'arable land', and 'áãñïíïìôá' (English equivalent: 'agronomy'): the management of land. 'Agriculture' relates literally to the cultivation and management of land, but in popular usage embraces virtually all patterns and scales of crop and livestock production, including horticulture, (literally: the cultivation of gardens: Latin *'hortus'* = garden), more generally used to designate the cultivation of fruits and vegetables. 'Agroforestry' or 'Agri-sylviculture' relate to the integration of crop and livestock production with natural and cultivated arboreal species, in some instances the production and utilisation of non-timber forest products such as tree fruits, nuts, wild honey and medicinal plants. 'Agribusiness' has come to embrace most industries that provide goods, machinery and services to agriculture, and the industries that preserve, process, store, distribute and market the products from agriculture.

Agriculture, horticulture and biotechnologies are predominantly devoted to human health: the production of food to provide energy for healthy activity, bodily growth and restoration, drugs to diagnose, prevent and cure diseases, and fibres from which garments and fabrics are created.

In recent literature, 'biotechnology' (singular) is frequently used exclusively to connote genetic modification of organisms and related modern biochemical reactions. In reality 'biotechnologies' (plural) embrace all processes that serve to protect, preserve, extract and convert materials of biological origin into products of industrial, commercial, social, economic or hygienic value and utility. Milling, baking and fermentation processes, devised by ancient Egyptians and Babylonians are as correctly designated 'biotechnologies' as are synthetic and extractive processes of biological materials that derive therapeutics and diagnostics from genetically modified organisms. One could cite instances of 'biotechnology' used to describe processes that have not progressed beyond a laboratory or field experiment, which more accurately should be designated 'biosciences'. In the subsequent text the plural noun 'biotechnologies' denotes the broader concept defined above.

EVOLUTION OF PLANET EARTH

The history of Planet Earth's existence can be classified geologically, according to the changing structure and composition of the earth's outer crust, or by progress in the development of tools, implements and technologies. Geological timing starts with the Precambrian period (about 4,500 million years ago), succeeded by the Cambrian (about 600 million years ago), which is believed to coincide with the appearance of the earliest fossils. Of particular interest to humans and their immediate ancestors is the Pleistocene period that began about 2 million BCE, during which period the Earth's climate underwent fluctuations between warm and cold temperatures, cycles that have continued until the present Holocene period that began around 15,000 BCE. The Holocene began with the end of the final glaciation, believed to have lasted from about 60,000 to 15,000 BCE. One of our many hominid ancestors, Neanderthal Man, a cave dweller and hunter, who appeared about 200,000 BCE, survived by hunting and gathering during the Pleistocene period.

From an agricultural perspective, an Earth-time classification based on the tools hominids used is perhaps more meaningful. It begins some 2.5 million years ago with the Early Palaeolithic period when hominids used primitive sharp-edged stone tools to harvest vegetation, cut up carcasses and sharpen sticks (Greek: Ðáëáéïò = Ancient; ëéøïò = stone). During the

Mesolithic period (Greek: ìåïò = middle or between) that began about 15,000 BCE, stone tools came with multiple sharp edges, attached to handles fashioned out of wood or deer antlers. The Mesolithic era merged into the Neolithic (Greek: íåïò = new) when more refined and efficient tools were made of polished stone, used as axe-heads to cut down trees, as sickles in crescent shapes with handles to harvest crops, with sharp blades to scrape animal hides. Harvested grains were ground by stone pestles and mortars, the pestle a heavy stone club, the mortar probably a concave hollow in a rock.

HUNTING AND GATHERING

Several scholarly narratives describe how our ancestors progressed from hunting and gathering to settled crop cultivation and animal husbandry [Blaxter 1992, Critchfield 1981]. Blaxter writes that to plan for the future we must build on knowledge accumulated from long past experience. The immediate past can provide a basis for simple extrapolation and adaptation of recent trends, while human history over the distant past can illustrate many of the difficulties that have confronted, and destructive practices that have constrained, humans in their quest for an adequate food supply. Blaxter discusses how changing weather patterns, massive epidemics, natural and man-made instabilities have been encountered and thus, in a broad sense, indicate the potential difficulties that lie ahead. Critchfield illustrates how settled agriculture, the practice of irrigation, domestication of animals and the invention of the plough led to the early establishment of village communities and their eventual expansion into towns and cities.

All animals in the wild survive by hunting and gathering, as, more than one million years ago, did the earliest hominids: the Australopithecines (African 'Ape-men'), the Pithecanthropoids distributed over Africa, Europe and Asia between 500,000 and 1,000,000 years ago. *Homo habilis* and the Neanderthals (*Homo neanderthalensis*) hunted and gathered from *ca* 200,000 to 35,000 BCE and devised Mousterian flint tools, more advanced and efficient than those of their predecessors. The first modern humans: *Homo sapiens* probably appeared between 10,000 and 15,000 years ago, being preceded by countless earlier primitive hominids. [McEvedy 1976;Weinberg 1994]

Anthropological opinions differ on when the various hominid species first appeared on earth and when some became extinct. *Homo erectus*, so named because of an apparent ability to walk upright, may have existed more than 1.5 million years ago. It is speculated that the Neanderthal hominids (*Homo neanderthalensis*) arrived about 200,000 years ago. Scientists at the University of Utah have studied evolution of certain head lice in tracing the history of early hominids. They conclude that *Homo neanderthalensis* and early *Homo sapiens* lived contemporaneously until about 25,000 years ago, when the Neanderthals and some other early hominids disappeared. The Utah scientists conclude that *Homo sapiens*, possessing larger brains and possibly bigger bodies may have destroyed their more primitive hominid cohabitants [Henderson 2004].

Hunting and gathering persists as a source of sustenance for many primitive people. Survival by hunting and gathering is sustainable if the dependent humans do not increase in numbers faster than the natural rate of food source regeneration. It is speculated that as each hunting-gathering tribe exhausted its immediate natural food resources it would migrate to find other sources of supply [Blaxter 1992]. Whenever a population grew faster than the natural regeneration of the wild animals and plants that provided its food some or all would die of starvation. It is not improbable, as has been the case in recent times, that the least productive, young children and old men and women, would be the first to be sacrificed in times of shortage.

EARLY HOMINID DIETS

The ways in which prehistoric hominids acquired their food and the composition of their various diets can only be conjectured, accurate determinations being virtually impossible. To begin, sites for suitable study must be found. In dry environments biological remains are better preserved, so more sites have been studied in arid and semi-arid ecologies than in the humid tropics. Samples suitable for analysis are plant and animal parts that have resisted decay for several thousand years. Fleshy, non-lignified plant parts rarely survive for any length of time. Carbonised seeds and highly lignified seed coats are among the best survivors. Bones of very small animals and of fish are less resistant to degradation than large animal bones and mollusc shells. Consequently, what can be analysed illustrates only an unknown proportion of the foods consumed by ancient hominids [Blaxter 1992; Verpmann 1973; Wheeler *et al* 1989].

Palaeoclimatology, carbon dating, pollen analyses, isotopic and elemental analyses of hominid skeletal remains are methods variously employed to discover some of the components of ancient hominid diets. The ratio of ^{13}C to ^{12}C in collagen from hominid bones indicates plant and animal sources that were dominant in the hominid's diet [Price 1989]. The ratio of ^{15}N to ^{14}N helps to distinguish between leguminous and non-leguminous diets. Where written or pictorial records are available, as from the early riverine civilisations of Egypt, Mesopotamia and the Indus valley, dates and patterns of the transition from hunting-gathering to settled farming can be more reliably determined. Nonetheless, present analytical methods cannot distinguish between plant and animal remains from a wild or a domesticated species. Evidence of lined storage pits and stone pestles and mortars do not irrefutably indicate that the grains stored and processed were cultivated, since both could be used for grains harvested from the wild. Out of the many animals slaughtered by hunters, only a few species appear to have been domesticated. Studies and samples analysed from the Mesolithic and early Neolithic periods show that wild red deer and roe deer were abundant, though neither seem to have been domesticated. [Blaxter 1992; Wheeler & Jones 1989; Clutton-Brock 1987].

Middle Palaeolithic hominids, whose remains are spread across Europe, West Asia, the Middle East and parts of Africa, were more successful hunters than their predecessors. Animal remains suggest they devoured many species of large mammals; how much of their meat came from hunting and killing live animals, how much from the scavenging of dead carcasses is unknown. Sites in Europe where bones from several animal species were found may have been primitive abattoirs where hominids processed the spoils of their kill. Archaeologists have discovered sharpened pointed stones probably used as spear heads, indicating a new kind of hunting technology. Simple hearths at many Middle Palaeolithic sites show that fire was used to cook meats, to dry and cook grains and, with protective clothing from animal hides, enabled Neanderthals to spread to and survive in such colder regions as the Central Asian steppes.

The Mesolithic period extended from the end of the Pleistocene ice age until the Neolithic, about 10,000 years ago, when settled farming first appeared in different locations and at different times.

The end of the Pleistocene ice age led to rapid environmental change where warmer post-glacial conditions of the Holocene epoch caused ice sheets to retreat and sea levels to rise. Temperate forests spread across Europe and Asia, large herds of mammals such as reindeer were replaced by red deer, roe deer and wild pig. Reindeer, elk and bison, adapted to cold climates, retreated to the north; the mammoth, giant deer and woolly rhinoceros became extinct.

FROM HUNTING TO SETTLED AGRICULTURE

Settled agriculture must be regarded as among the most important innovations and developments in human history. For whatever reasons particular communities changed from hunting-gathering to agricultural technologies devised and developed from perceptive empiricism, each adopted system being influenced by local soil, climate and resource conditions. There seems general agreement among archaeologists and ethno-botanists that transitions from hunting and gathering to settled farming occurred in different ways, according to different patterns and stimuli in different places. Settled agriculture, in which food-crop seeds were planted and animals domesticated began during the Neolithic period around 10,000 years ago. In some communities settled farming seemed to result from local indigenous development, in others farming technologies devised or discovered in one location radiated outwards to be adopted or adapted by other communities [Blaxter 1992, Critchfield 1982].

To evolve from gathering to crop cultivation required the discovery of indehiscent plant species. In order to reproduce themselves, wild grasses and other plants when mature scatter their seeds, some of which will then take root. Seed scattering plants are described as dehiscent (= bursting or splitting open). This trait of dehiscence is particularly noticeable in sesame, hence the expression 'Open sesame'. Settled cultivators have need of indehiscent plants so that, when mature, the seeds can be harvested and stored, to be eaten or later planted.

The earliest farmers were of Neolithic culture located in the region now covered by Iraq, Iran, Jordan, Syria and Turkey, along the Nile in Egypt, the Indus river valley and its tributaries in what is now India and Pakistan. Exactly when the first crops were derived from planted seed is uncertain and varied among regions. Though most are believed to predate 6000 BCE, some ethno-botanists believe the earliest may date from around 10,000 BCE. According to carbon isotope dating, wheat, barley and possibly flax were cultivated in Mesopotamia from 8,000 BCE. Wheat and barley cultivation slowly spread northwest to reach Britain about 2,000 BCE. A diversity of fruits and vegetables: onions, melons, cucumbers, dates, figs were cultivated in Mesopotamia and nearby regions between the fourth and third millennia. During the second millennium, cotton was grown and spun in India and Egypt, linen and silk in China.

Examination of animal and plant remains by 14C indicates that sheep, bovines, goats, pigs and asses were domesticated in Mesopotamia between 9,000 and 6,000 BCE. Neolithic settlements were more permanent than the camps of nomadic hunters and gatherers though some early farming settlements had to move periodically if weather patterns changed or soils became infertile from excessive cultivation [Blaxter 1992].

The first crop cultivation systems probably began by clearing land for continuous cropping until soil fertility was exhausted, at which time the cultivators would move to another location, which then and now in certain poor nations is the basis of slash-and-burn agriculture. It is generally assumed that the early cultivators started by empirical discovery of which wild plants were edible and from which seed could be removed and planted on cleared land. From the harvest of their cultivated crop, farmers learned to save some seed for the next planting. Repeated cultivation of a chosen genotype would eventually provide the farmer with a stable crop with relatively homogeneous characteristics. Herds of goats and sheep were assembled, captured young wild animals with the most useful characteristics, those with small horns and high milk yields, being domesticated. Their progeny would become progressively tame, with comparatively uniform physiologies.

DEMOGRAPHIC, ECONOMIC AND SOCIAL CHANGE: THE RIVER VALLEY CIVILISATIONS

Settled farming, with domestication of plants and animals led to profound social changes, particularly in regions such as Mesopotamia, Egypt and the Indian Harappan culture, where fertile soil, ample water for crop irrigation and abundant sunlight provided conditions for highly productive farming systems. Archaeologists have located early farming sites in Thailand and China, where rice and other indigenous crops were cultivated, in a few African territories where indigenous sorghum and millets were grown. Most comprehensively described were farm settlements along the Euphrates, Tigris, Nile and Indus valleys [Critchfield 1981; Thapar 1996].

The capacity to produce food surplus to the producers' needs, facilitated trade in agricultural commodities, the establishment and growth of non-farming urban communities. It also led to income stratification as prosperous farmers and urban traders increased their wealth. Cultivators in Mesopotamia and Egypt were probably the first to use systematic irrigation, to establish cities, towns and regulated social organisations. The Indian civilisations were not far behind. The development of cuneiform script in Mesopotamia and Egyptian hieroglyphics [the writing of priests] provided means of communication and records for posterity [Leick 2002; de Burgh 1947].

Mesopotamia

'Mesopotamia' means 'between rivers' the rivers being the Tigris and Euphrates that rise in the Anatolian mountains and flow south to the Persian Gulf. In contemporary geographical usage, Mesopotamia includes Iraq, eastern Syria, parts of Iran and much of south-eastern Turkey. It comprised two distinct ecological zones: the northern Fertile Crescent, where rain-fed and irrigated agriculture started and the first urban settlements appeared some 10,000 years ago; and the southern flood plain, the rich fertile land where, in dry seasons, crops were irrigated from an ingenious system of dykes and canals. The most enduring legacies of ancient Mesopotamia are the dozens of established cities, and the invention of writing. Each of the cities strung along the rivers controlled its own rural territories, its agricultural production and system of irrigation, its individual form of governance and administration.

Many of the ten principal cities – starting with Eridu about 8,000 years ago and on to Babylon, which around 2500 BCE grew from a small town to a magnificent city that lasted into the first centuries of the contemporary era – disappeared over time variously because of climatic change, rivers changing their courses, military conflict and reasons not fully understood. Most now exist only as archaeological remains [Leick 2002].

According to some historians, most of Mesopotamia was under water prior to the Chalcolithic period. This is possibly the origin of the Biblical flood mythology. There is an ancient Sumerian myth that the storm god Enlil decided to reduce the population by drowning them in a flood. The god of wisdom, Enki, persuaded a wise man Atrahasis (also known as Ziusudra) to build a boat and to stock it with seeds and species of all living organisms, so as to ensure the survival of all creatures when the flood subsided [Armstrong, K 2005].

As the climate gradually changed, some 6,000 years ago, the floods receded, and the plains became habitable. The post-deluvian virgin alluvial soil supported barley, wheat, flax, horticultural crops and livestock production. The southern marshlands were a substantial source of fish, game and wild life. With water from the Tigris, Euphrates and their tributaries, crops were irrigated during seasons of low rainfall. Thus, starting in the fourth millennium, 4000 to 3000 BCE, southern Mesopotamia [Babylonia, Akkad and Sumer] provided an extensive exploitable ecosystem able to produce abundant food, thus encouraging settlements which expanded into large urban areas. Produce from farmers, fisher-folk and hunters stimulated a transition from subsistence agriculture to rural-urban complexes. Donkeys and oxen were probably the main beasts of burden.

Urban Commerce

The urban traders, artisans and administrators were fed by abundant food from the surrounding cultivated land. Artisans developed pottery from native clays, textiles out of flax and wool of domesticated sheep. Administrators prescribed and enforced rules that restrained conflict, enabling people spread among the settlements to live together in comparative peace. Surplus grain, leather from sheep hides, sun-dried fish, dates, other fruits, textiles and timber from natural forests, and locally mined metals and semi-precious stones were exported to urban settlements along the navigable Tigris and Euphrates [Leick 2002].

Theophrastus, Aristotle's celebrated pupil, in his *History of Plants*, describes how in ancient Babylon wheat fields were harvested twice each year, beasts being subsequently fed on the residual stubble. He stated that on cultivated, irrigated land each planted wheat seed could produce between a 50 and 100-fold harvest. One assumes he was comparing the total weight of biomass harvested not the weight of harvested seed to the weight of the seed planted. On fertile soil irrigated from nearby rivers, Babylonian farmers could harvest two crops every year, sufficient to satisfy their own needs and those of the expanding urban populations. [de Burgh 1947].

Egypt and the Nile

The Greek historian Herodotus wrote: "Egypt is the gods' gift of the Nile". Through Egypt, the Nile flows some 1,200 km from the first cataract in the north to the broad delta where it empties into the Mediterranean. The Egyptian Nile is fed from the White Nile, which originates in northern Uganda, supplemented by the Blue Nile, which flows in at Khartoum, and the Atbara which joins about 450 km north of Khartoum.

Both the Blue Nile and the Atbara originate in the Ethiopian highlands. Over many centuries the abundance of the Nile, the fertile silt deposited after each annual flood, and Egyptians skilful management of dykes and irrigation systems supported a flourishing agrarian society. Diverse plant species were cultivated for food, clothing and shelter, well into the period when, under Roman rule, Egypt was regarded as the breadbasket of the Mediterranean [de Burgh 1947; Livy 1960; Cowell 1948]. The Nile sustained Egypt's fishing industry for several thousand years.

The Egyptian population grew from several hundred thousand during the pre-dynastic period (5000–3000 BCE) to several million by 1000 BCE. Egypt's cities and settlements were located close to the Nile, the densest population being along the flood plains. During most years when the rains in Uganda and the Ethiopian highlands were abundant, the Nile overflowed its banks. The floodwaters as they receded deposited a rich, black fertile soil to sustain productive crop cultivations over many millennia [McEvedy 1976].

History records the decline and fall of many once prosperous and advanced nations. Egyptians were literate, numerate and practised productive irrigated agriculture more than 5,000 years ago. They devised a decimal numerical system and the means to communicate in writing. The mathematics of dynastic Egypt was significantly influenced by agriculture. Calculations of area, length × breadth, were used to determine the size of farmers' fields, and before the fourth Millennium Egyptians calculated the value of ∂ [pi] the ratio of circumference to diameter as equal to 3.145. The present accepted value is 3.143. Knowing the value of ∂ enabled Egyptians to calculate the storage capacity of cylindrical grain bins.

Indus and Harappan Culture

The earliest traces of hominid activity in India go back to the second interglacial period between 400,000 and 200,000 BCE and show evidence of stone tools. There followed a long period of slow evolution that culminated in the spectacular Indus valley civilisation, known as the Harappan culture, which matured around 2500 BCE. The antecedents of the Harappan culture were village sites in the Baluchistan hills and rural communities along rivers that flowed in Rajasthan and Punjab. The Vedic civilisation, which began in the Indus valley, is said to have believed more in monuments of the mind and thought than in material monuments. Archaeological excavations have brought to light some 80 settlement sites dating from about 3500 BCE. The Indus civilisation was among the world's earliest; Indian historians wisely state that civilisation began when cereals and cattle were first domesticated. The Indus civilisation culminated in two main centres: Mohenjodaro close to the Indus valley, and Harappa on the Ravi river. Wheat cultivated near to Mohenjodaro is claimed to be an ancestor of the wheat varieties now grown in the Punjab. Both Mohenjodaro and Harappa maintained large granaries in which the annual wheat harvest was stored. In India cotton was first grown. Grain, cotton and other commodities were transported in wheeled carts drawn by oxen, and by river boats [Mookerji 1961].

The Harappan culture was essentially urban, the two main cities Mohenjodaro and Harappa being maintained from surplus crops produced by the surrounding rural farmers and stored in elaborate granaries. Harappan communities developed a flourishing trade with people in the Persian Gulf and Mesopotamia [Thapar 1966]. Some historians believe the first Sumerians to settle in Mesopotamia migrated from Baluchistan [Bouquet 1962].

Roman Agriculture

Marcus Tullius Cicero, Roman Consul, politician and philosopher, wrote: "Of all gainful occupations nothing is more pleasing and better becomes a well-bred man than agriculture" [Cowell 1948]. Some 6,000 years ago active volcanoes in Central Italy covered the Latin plain with ash and lava that provided Roman farmers with an extremely fertile soil. For several centuries Roman farmers enjoyed exceptional prosperity. Between 265 and 70 BCE the number of male Roman citizens rose from 290,000 to 900,000. When women, slaves and non-citizens were counted, during the last century BCE, the total Roman population probably exceeded 3 million [Tacitus 1965].

But soil fertility and productivity could not keep pace with population growth, a constraint aggravated by top-soil stripped of its protective tree and grass cover being washed away by torrential rains. Animal manure and wood ash were insufficient to replace lost soil nutrients and what was once fertile crop land degenerated into land fit only for grazing animals. To pursue Rome's military adventures and conquests, many of the best farmers were recruited into the Roman legions. By the turn of the millennium, as BCE changed to CE, Rome imported about 190,000 tonnes of wheat from Sicily, Sardinia and North Africa. Soon thereafter Egypt, the principal granary of the Mediterranean, became the main supplier of food grains to Republican Rome [Cowell 1956; Livy 1960].

Of the three civilisations whose early prosperity was derived from productive agriculture, only India remains as an efficient producer of food grains. As recorded in *Indian Agriculture 2003*, published by the Indian Economic Data Research Centre, since the 1950s India's annual wheat harvest has risen from 6 million to 70 million tonnes.

From Manual to Mechanised Processes

Virtually all the technologies employed to provide for basic human needs – agriculture, food processing, drugs, pottery and textiles – were discovered and developed empirically. Domestic and artisanal technologies were practised for many centuries before there was any scientific understanding of why they worked. Scientific understanding began in the late eighteenth and nineteenth centuries; only during the last century have novel biotechnologies been devised and developed based on scientific principles. For many centuries most apparent progress was from gradual, episodic mechanisation: the replacement of human hands by machines. Long ago iron replaced wood in simple ploughs, the primitive iron plough, stronger than wood, turned top-soil over rather than just scratching the surface. Kates [1994] illustrates how new tools and technologies have stimulated human population growth over the past million years by raising available food supplies. Progressive improvement in hand tools used by hunters and gatherers resulted in an estimated population increase from 150,000 to

5 million during the period from one million to 100,000 years ago; a second surge from 5 million to *ca* 1 billion came with expansion of settled agriculture between *ca* 8000 BCE and the middle of the nineteenth century, since when the application of science to food production, preservation, processing and distribution has stimulated a world population explosion to more than 6 billion [Kates 1994; Hulse 1995].

Most rapid and far-reaching farm mechanisation spread across the Western Great Plains of the United States during the nineteenth century. The steel blade ploughshare appeared in 1830, the reaper in 1834, soon followed by mechanised harrows, seeders, threshers and binders, many manufactured and marketed by Cyrus McCormick. Mechanisation disruptively reduced the need for human labour on American farms: one McCormick threshing machine operated by four men replaced 400 farm labourers. Towards the end of the nineteenth century, guano from Peru and nitrates from Chile began to replace or supplement animal manure as nitrogen fertilisers. Introduction of the steam tractor accelerated farm mechanisation. During the second half of the nineteenth century more than 160 million hectares of new land came under cultivation in the United States [Nye and Morpurgo 1955].

As in ancient Egypt and Mesopotamia, rising farm productivity stimulated the growth of cities, the rate of American city expansion being unprecedented in human history. In 1910 Chicago's population reached 20 times the size in 1860. Mechanisation, more abundant farm productivity and rural-to-urban migration occurred in Europe and North America. During the twentieth century, the proportion of the Canadian work force employed on farms fell from *ca* 40 per cent to barely 3 per cent. The human misery suffered in cities that grew too rapidly, without planning or provision of essential services, is forgotten or ignored among nations of Africa, Asia and Latin America where urban expansion is uncontrolled, where slums and unhygienic environments are all too familiar.

From Empiricism to Scientific Agriculture

One of India's most distinguished scientists, the late Professor Y. Nayudamma, made the profound observation that science and technologies serve no useful purpose if they fail to provide for the needs of human beings. India's agricultural scientists and farmers in productive States have served its peoples exceptionally well, transforming the nation from dependence on imported grains and food aid to an abundant self-sufficiency in food grains.

Contrary to long-standing belief, agriculture does not inevitably progress starting with scientific research that begets invention, that leads to technological innovation, and ultimately results in ever-increasing crop production. Farming systems research (FSR), that starts by assessing and understanding farmers' adopted technologies, opportunities, constraints, resources and risks, began during the early 1970s. The benefits to farmers realised by FSR have been less widely publicised than research that generated higher yielding cereal crop phenotypes. It is through modifications based on careful systematic analysis and understanding of existing farming practices that sustainable systems are most effectively and sustainably developed, adapted and adopted. [IDRC 1981; Zandstra *et al* 1981].

Blaxter contends that modern food systems, highly dependent on energy other than solar radiations, are inherently unsustainable. As has come about over the past 12,000 years, novel or improved technologies and systems are needed to maintain an equitable, sustainable balance between food supply and persistently rising demand [Blaxter 1992]. Blaxter estimates that roughly 98 per cent of today's world's food supply is derived from terrestrial ecosystems. Though some 80,000 identified plants and countless animal species are believed to be edible, roughly 90 per cent of all human food calories come from only 15 plants and eight animal species.

Early Scientific Evolution

In his imaginative novel *Brave New World* Aldous Huxley asserted that the world's history of economic, technological and industrial development can be divided into two periods: a. before and b. after Henry Ford, the Detroit automobile industrialist who devised the continuous on-line process by which to assemble automobiles. In this author's opinion a more rational division is pre-and post-Faraday. If the Uruk-Sumerian invention of cuneiform script in the fourth Millennium, and Johannes Gutenberg's creation of the printing press in the 1450s can be considered the most important contributions to international scholarship, it seems appropriate that Michael Faraday started his working life as a bookbinder. Though he never graduated from a university, Faraday's fascination with, and self-acquired knowledge of science led to his appointment as scientific assistant to Sir Humphrey Davy. Faraday's renowned contribution to electromagnetism began when he reviewed the Dane Hans Christian Oersted's discovery of the magnetic effect associated with an electric current. Faraday was the first person to discover benzene whose six-carbon ring structure, later elucidated by Kekule, is of fundamental importance to organic chemistry, to agricultural and food biochemistry [Schumm 1987].

Science Together with Empiricism

Although during the fourth century BCE Aristotle classified more than 500 animal species within hierarchies, and his pupil Theophrastus wrote a *History of Plants*, agricultural practices changed little from the time men and women began to cultivate crops and husband livestock around 10,000 years ago. Thales, one of the first Greek philosophers in the sixth century BCE, postulated that water was the main constituent of all plants. But it was not until the seventeenth century CE when van Helmont observed and reported the dependence on water of his experimental willow, Stephen Hales demonstrated how water is transmitted through roots and leaves, Joseph Priestley and Jan Ingenhosz described leaf respiration, and von Liebig and Dumas experimented with plant uptake of nitrogen and other nutrients, that an understanding of what influences plant growth began to emerge [Mees 1947; Gribbin 2002].

From the time of Faraday and his contemporaries in the nineteenth century scientific explanations of natural phenomena, of technologies empirically discovered, devised and developed over many millennia, gradually emerged. Nonetheless, while scientific underpinnings were coming to light, discovery and progress from thoughtful observation was widely evident. During the eighteenth century an English farmer, Coke of Okeham, improved soil fertility with little help from chemists; while Robert Bakewell increased meat and wool per animal by selection and breeding of Leicestershire sheep more than 100 years before Mendel discovered the laws of heredity. The famous Rothamstead experimental station blossomed out of field experiments started by a private farmer John Lawes, who bore the total cost for almost 60 years until in 1889 a sustaining Trust Fund was endowed to carry on some of the world's longest systematic studies of soil fertility. During the seventeenth century, Grew, Camerarius and Sprengel described plant male and female organs – stamens, styles, ovaries and ovules – and how insects carried pollen between plants.

Microscopy and Cytology

Who first discovered the magnifying power of convex lenses is unknown. The thirteenth century English scholar Roger Bacon is said to have constructed eye spectacles. During the seventeenth century Galileo devised a compound microscope, equipped with convex lenses he had ground. In 1625 he published microscopic details of various insects. In the late seventeenth century, a self-educated Dutch microscopist Anton van Leeuwenhoek, who earned his living in a drapery store, delicately ground glass lenses used to study and describe previously invisible microscopic materials. He described tiny parasites that infested fleas, a discovery that inspired Jonathan Swift to write:

So naturalists observe, a flea

Has smaller fleas that on him prey;

These have smaller ones to bite 'em

And so proceed ad infinitum.

Van Leeuwenhoek described other cellular organisms that he named 'animalcules', which probably were protozoa or bacteria. During the second half of the nineteenth century, bacteria were more accurately identified by Louis Pasteur, who demonstrated how micro-organisms could be thermally inactivated.

From Alchemy to Chemistry

Robert Boyle, son of the Earl of Cork, is most renowned for demonstrating the inverse relation between gaseous volume and pressure. Whether or not his book, *The Sceptical Chymist: Chymic-physical doubts and paradoxes*, provided the bridge between alchemy and chemistry, Boyle undoubtedly pioneered scientific method, inspired by Francis Bacon's thesis that scientific investigations must begin by critical assessment of all relevant available data, not by imagining a wonderful idea then seeking facts to sustain it. From the time of Faraday, progress in scientific method, in analytical chemistry, microscopy and cytology progressed to a greater understanding of living organisms, development of agricultural, food and pharmaceutical technologies.

The quarter-century following Faraday's empirical discoveries of electro-magnetic induction and the anaesthetising properties of ether was a vintage period in scientific progress. Darwin published *The Origin of Species*, Mendel formulated his laws of genetic inheritance, Maxwell published his theory of electromagnetism, Pasteur and Koch described the microbial origins of many diseases, Kekule elucidated the cyclical structure of benzene, Gibbs published his first laws of thermodynamics related to chemical reactions, Geissler's vacuum-tube technology paved the way for discovery of electrons ['cathode rays'], X-rays and the eventual discovery of radioactivity. A congress at Karlsruhe, convened by Kekule, led to a rational nomenclature for chemical compounds. The classification of elements according to their atomic weights by Cannizzaro, Newlands, de Chancourtoix and Meyer culminated in Mendeleyev's periodic table of all known elements. The later classification by atomic number did not greatly alter the Medeleyev table.

Joseph Black, Henry Cavendish, Joseph Priestley and Carl Scheele advanced the development of chemistry as a distinct science. Their discoveries of carbon dioxide, oxygen, hydrogen, ammonia, hydrogen chloride and sulphur dioxide; of the composition of water; and their finding that air is a mixture of gases, are illustrative of what was achieved by men of immensely diverse interests, most of whom, by contemporary scientific standards, would be classed as non-professionals [Gribbin 2002]. An attempted dissertation of all that science and technology have contributed to modern agriculture and biotechnologies lies far beyond the scope of this publication. Nevertheless, a few exceptional persons deserve special mention.

Lavoisier

Antoine-Laurent Lavoisier was born in 1743 in Paris, the son of a well-to-do lawyer. Lavoisier qualified as a Licenciate of Law, at the same time studying astronomy, botany, chemistry, geology and mathematics. He never practised law, his first work being a project to complete a geological map of France. During the late eighteenth century, Lavoisier devoted his talents to chemistry, demonstrating that when sulphur burns it gains weight, the first step in understanding how atmospheric oxygen (Priestley's 'pure air') is critical to combustion processes. From later experiments he concluded that animals derive their body heat from combustion processes dependent on inspired oxygen. He determined the weight of ice melted during ten hours by the heat from a guinea pig's body. Then he measured the weight of carbon burned to melt the same weight of ice. He measured how much 'fixed air' or CO_2 the guinea pig exhaled during ten hours and how much 'fixed air' was generated by burning different weights of charcoal. Lavoisier thus anticipated von Liebig's demonstration, 100 years later, that animals derive their body heat and energy from *in vivo* conversion of carbohydrates and lipids.

Lavoisier was the first to demonstrate that water is formed by the combination of hydrogen ('inflammable air') with oxygen, and how 'fixed air' is a compound of carbon and oxygen. His books *Traité elementaire de chimie* and *Méthode de nomenclature chimique*, published in 1787, laid the foundation for chemistry as a distinct scientific discipline, and provided the basis for a logical and systematic nomenclature for chemical elements and compounds: 'oxygen' replacing the earlier 'phlogisticated air', 'inflammable gas' giving way to 'hydrogen', 'oil of vitriol' being renamed sulphuric acid. One-third of the latter book was a dictionary of new chemical names presented in alphabetical order together with the old names, many coined by alchemists on the basis of the colour and physical appearance of the substance [Gribbin 2002; Crosland 1978; Ward and Dubos 1972].

Lavoisier's many profound pronouncements include: "La vie est une fonction chimique", demonstrably true in that all life processes are sustained by the conversion of one form of energy into another by biochemistry; and "Scientists should be as precise and meticulous in their written and spoken words as in their experimental measurements", advice all too often ignored in speeches and writings about 'sustainable development'. Due to an earlier involvement with Tax Farmers, in 1794 the French revolutionaries decreed that Lavoisier be executed by the guillotine.

Chemical Nomenclature

In 1806 Berzelius defined Organic Chemistry as: "That which describes the composition of living bodies and the chemical processes within them", the first distinction being between substances of vegetable and animal origin. Benefiting from Lavoisier's earlier publications, in 1860 Kekulé convened a congress in Karlsruhe at which a group of distinguished scientists discussed systems of nomenclature and symbols for organic compounds. From this congress and subsequent discussions Organic Chemistry progressed from Berzelius' definition to the present concept: the complex chemistry of thousands of carbon compounds [Crosland 1978; Dixon 1973]. It was not until the twentieth century that Berzelius' concept of organic chemistry evolved into the discrete discipline 'biochemistry', at first designated 'physiological chemistry'. From a typical dictionary definition: "Biochemistry is the study of the chemical constituents of living organisms, their functions, reactions and transformations" [Rose 1970].

TAXONOMIC CLASSIFICATION

Ray and Linnaeus

Arguably, the most productive naturalist of the seventeenth century was an Englishman, John Ray. Ray travelled extensively throughout Europe observing, describing and collecting thousands of specimens of natural flora and fauna. The results of his studies are reported in his books: *Ornithology*, *History of fishes* and *History of plants*. The latter, published in three volumes between 1686 and 1704, describes more than 18,000 plants classified according to distribution, natural habitat and those believed to possess pharmacological properties. Ray established 'species' as the basic units of taxonomic classification. His book "History of insects" was published posthumously.

Ray's taxonomy was based on anatomy, morphology, physiology and systematic principles adopted by the better known Carolus Linnaeus, born in 1707 in Sweden [Gribbin 2002]. The Linnaeus family name, Ingemarsson, was legally changed by his father to Linnaeus, the Latinised version of the Linden tree. Though he graduated in medicine, Linnaeus' life-long interest was in biology. He noted differences between the reproductive organs of flowering plants as a criterion of classification. He lived until 1778, switching from the Chair of Medicine to the Chair of Botany at Uppsala University in 1742. His most famous book, *Systema natural*, was first published in 1740. His binomial system of naming species was spelled out in Volume 1 of the tenth edition in 1758. Linnaeus classified close to 8,000 plant species and some 4,400 animals, all arranged in a hierarchy from 'Kingdom' at the top to 'species' at the bottom. As taxonomic techniques were refined, some of Linnaeus' original Latin binomials have changed; but his principles remain as the standard of species identification: a Latinised generic plus a specific name: e.g., *Canis lupus*, the scientific name for 'wolf'.

Linnaeus was the first to include humans in a zoological classification system. He questioned why humans are the sole species '*sapiens*' in the genus '*Homo*'. In his 1746 publication *Fauna Svecica* he writes: "I have yet to find any characteristics by which man can be distinguished on scientific principles from an ape." Based on modern DNA evidence man should be classified among the chimpanzees, possibly as '*Pan sapiens*'. The accepted classification '*Homo sapiens*' is used so as not to offend certain religious sensitivities [Ward and Dubos 1972; Gribbin 2002]. As modified by later taxonomists Linnaeus classification of animals now reads:

Kingdom	*Animalia*
Phylum	*Chordata*
Subphylum	*Vertebrata*
Class	*Mammalia*
Order	*Primates*
Family	*Hominidae*
Genus	*Homo*
Species	*Sapiens*

After Linnaeus died, James Smith, a wealthy English botanist, was instrumental in 1788 in founding the Linnean Society which, in London, holds most of Linnaeus' collections of biological specimens [Gribbin 2002].

Genetic Inheritance

The years in the nineteenth century when Faraday lived (he died in 1867) yielded an abundance of discoveries that advanced the development of agricultural science. In 1837, Schleiden proposed that all plant tissues are composed of cells. A year later, Schwann suggested that all living organisms are of cellular composition. In 1858 Virchow, a professor of pathology in Berlin, asserted that all cells are derived by division of pre-existent cells. Using more powerful microscopes, Hertwig and others later discovered cell nuclei and, working with sea-urchins, observed how sperm penetrated into eggs, the nuclei of the sperm and egg cells fusing into a new nucleus that combined material from both parent cells. Flemming and van Beneden reported the presence of thread-like structures in cell nuclei, which they named chromosomes because of their susceptibility to staining by chemical dyes. At Freiburg, Weismann described how when a cell divides its chromosomes are duplicated to be shared between the two daughter cells. Weismann declared chromosomes to be carriers of hereditary characteristics between parents and progeny (Hulse 2004).

Mendel

Now recognised as the founder of classical plant breeding, Mendel was born in 1822 into a farming family in Moravia (then in the Austrian empire). Being too poor to afford a university education, Mendel joined the priesthood where he changed his Christian name from Johann to Gregor. After completing his theological studies, recognising Mendel's exceptional intellect and intelligence, the abbot of his monastery in Brünn (now Brno in the Czech Republic) enabled Mendel to study at the University of Vienna. There he acquired a knowledge of physics, chemistry, statistics, probability mathematics and plant physiology.

After two years, and without a degree, Mendel returned to the monastery where, in 1856, he began a ten-year study of heredity in peas. Working with several thousand plants, on a plot 35 × 7 metres, Mendel transferred pollen by hand from one plant to another, meticulously recording his every action and the results. His results from hundreds of repeated experiments, all subjected to rigorous statistical methods learned in Vienna, revealed that when peas with rough wrinkled seed coats were crossed with smooth-coated seeds, 75 per cent of the progeny had smooth coats, 25 per cent were rough, wrinkled seeds. This phenotypic phenomenon of unequal inheritance of parental characteristics illustrated the now familiar principle of dominant and recessive genes. Contrary to Darwin's theory of heredity, Mendel demonstrated that inheritance results not in a uniform blending of parental characteristics but from specific traits being inherited in unequal proportions.

Mendel's unique results were overlooked for more than half a century, in large part because in 1868 his abbot died and Mendel was appointed as the abbot's successor. His religious and administrative responsibilities allowed time neither to continue his plant breeding experiments nor to promulgate and disseminate his results to many who could have benefited. The rediscovery early in the twentieth century of Mendelian laws of inheritance, together with the elucidation of chromosomal behaviour, laid the foundations for highly productive plant breeding, most notably the so-called Green Revolution during the latter half of the twentieth century. [Stern and Sherwood 1966; Stoskopf 1993; Wortman and Cummings 1978; Gribbin 2002].

Closing Comment

The foregoing illustrates how biotechnologies were discovered, devised and developed over many millennia; how the basic biotechnologies – agriculture, food, drugs and textiles – began as domestic or artisanal crafts and gradually, over many millennia, were modified by perceptive empiricism. Early, primitive technologies preceded the most rudimentary scientific understanding by many hundreds of years, systematic scientific discovery being of very recent origin with much yet to be learned. What is known of the development of biotechnologies essential to human health and survival: agriculture, food, drugs and textiles, can truly be described as 'a gradual unfolding' and illustrate that sustainable progress can rarely be hastened but more reliably results from persistent patient systematic imaginative activity.

As is emphasised later, all development – to be rational, productive and sustainable – requires a systematic understanding and logical integration of all that influences and impinges upon the intended development. This is especially true of development related to the biotechnologies, upon which human life, biodiversity and their environments depend, biotechnologies which are progressing and changing more rapidly than at any time in the past.

POVERTY

Poverty and Development

Few persons and agencies dedicated to and experienced in economic and social development would disagree that poverty and its alleviation is a fundamental purpose of all development programmes. Different dictionaries and social commentaries define 'poverty': 'the condition of owning neither wealth nor material possessions'; 'having insufficient income and resources to sustain healthy survival'; 'suffering a lack of basic human needs: adequate food, clothing, housing, clean water, health services and hygienic sanitation'. These quotations define 'absolute' or 'near-absolute' poverty and destitution. Except for a few extreme cases, among affluent nations poverty exists as 'relative' rather than as 'absolute' poverty, where certain segments of society own less and receive lower incomes than the majority. The world's poorest people, most in Africa and Asia, are so poor they struggle to survive.

Defining Poverty

To determine and agree upon a precise, internationally recognised definition of 'poverty' is a difficulty that confronts all development, government, non-government and social service agencies. If accepted that those classed as LICs are the poorest nations, in 1982 and 1995 respectively, people with less than $410 and $765 per cap/year would be considered below the poverty line. An enquiry addressed to several international development agencies in early 2004 revealed that, for some the poverty line is set at $1.00/Cap/day ($365/Cap/year) for others $2.00/Cap/day ($730/Cap/year). Several United Nations agencies now regard those whose income is less than $1.00 USD per day as poverty-stricken.

In the United States the 'Official Poverty Line' is an income level determined by the Bureau of Census: poverty classifications among US citizens depend upon in what degree their-estimated incomes fall below the Official Poverty Line. In Canada, definitions of what constitutes the poverty line vary among government, non-government and welfare agencies. By some agencies the poverty line is set at the estimated median income for all Canadians. According to the Canadian Council on Social Development, the poverty line will vary according to family size, the location and population dimensions of the community in which the family lives.

In the year 2002, the poverty line for a family of four persons could range from $36,250 CAD ($30,000 USD) to $25,000 CAD ($21,250 USD), dependent on whether they lived in a large city, a medium town or a rural community. It is apparent that those classed below the poverty line in Canada suffer far less abject deprivation than Africans and Asians who fall below defined poverty lines of $2.00 or $1.00/Cap/day. The poor of Africa and South Asia, unlike Canadians, do not enjoy access to beneficial health and social services. Millions of poverty-stricken Asians and Africans would be delighted to enjoy the standard of living, social and health services available to many Canadians classed as below the poverty line.

The Poverty Literature

Canadians were awakened to the existence of widespread abject poverty by two series of broadcast lectures during the 1960s: in 1961 by Barbara Ward [Ward 1961] and in 1965 by John Kenneth Galbraith [Galbraith 1965]. The dismal situation and the means of alleviation cited by Ward and Galbraith are re-echoed in several of Professor M.S. Swaminathan's perceptive publications [Swaminathan 1999, 1999a, 2002]. The poverty described by Ward and Galbraith 40 years ago remains all too pervasive in 2006.

Galbraith States: "The first and most obvious consequence of poverty is social and civil instability", a truism woefully ignored by contemporary political leaders. Galbraith vividly describes how poverty pursues a downward cycle: poverty begets more poverty since the poor cannot take advantage of remedial opportunities or adapt productive innovations that incur adverse risks. In contrast, wealth begets more wealth and enhances the power of the wealthy to exploit those too poor to protect themselves.

Barbara Ward, in her brilliant historical review of human progress, her Massey Lectures on *The Rich Nations and the Poor Nations,* defines the political, socio-economic, biological and technical actions required to alleviate poverty:

a. Equal opportunity and equitable access to resources among people and nations without distinction of class, race, religion or sex.

b. Universal access to resources and services essential to survival, health and security.

c. Food production and distribution must keep pace with population increase and changes in demographic distribution.

d. Transformation from artisanal crafts to industrial technologies; encouragement and support for rural employment in agribusiness; technological development relevant and appropriate to the needs, resources and opportunities of those who are to apply and benefit from them.

It is often difficult to differentiate between causes and consequences of poverty, in many instances the two being mutually reinforcing, as described by Galbraith. Chronic malnutrition impairs the ability to learn and to work, which in turn deprives those so afflicted of opportunities for stable, gainful employment. Histories record how relative poverty can degenerate into absolute poverty when those with wealth and power oppress and enslave those too weak to resist. Among several ancient civilisations, enlightened rulers, such as Hammurabi, enacted laws to protect the poor from predations by the rich and powerful [Harper 1999].

Extensive discussions of poverty, its attributes, causes and consequences, are published by the International Development department and the Institute of Poverty Research at the University of Manchester in England.

Magnitude and Extent of Poverty

The UN *Human Development Report* for 2002 estimates that at least 1.2 billion – 20 per cent of total world population – struggle to survive on less than $1.00/day. During the 1990s, if one excludes the Peoples' Republic of China, the numbers in the less than $1.00/day category increased by 28 million. Only 30 of the 155 countries classed as LICs or nations in transition from 'developing' realised *per caput* income growth of more than 3 per cent/yr. The following two tabulations show per cent population with less than $1.00/day (Table 7); relevant regional data during the 1990s (Table 8):

Per cent population below $1.00/day

Sub-Saharan Africa	49
Nepal	38
Nigeria	70
Bangladesh	36
Ethiopia	80
India	35

In Europe the first organisations to provide succour to the poor and destitute were the medieval monasteries. In 1597 the English Parliament enacted a Poor Law that elaborated on earlier Elizabethan legislation to assist the indigent, including a compulsory Poor Rate in 1563, and the method by which to collect it in 1573. During the sixteenth and seventeenth centuries town councils in England were required to provide for various forms of social distress: homes and schools for children, hospitals for the sick, relief for the aged, work places for those willing and able to work, correctional institutions for people classed as deliberately idle [Bindoff 1950]. The 1795 Speenhamland Act revised English relief to the poor by delegating to local Parish Councils the responsibility for supplementing rural labourers' wages to a subsistence level.

Income growth and poverty reduction 1990s

	Income growth Avge/%/cap/year	**Poverty reduction %**	**Number [M] below $1.00/day**		**% below $1.00/day**	
			1990	**1999**	**1990**	**1999**
S S Africa	—[0.4]	—[1.6]	240	315	47	49
E Asia/						
Pacific	+ 6.4	+ 14.9	485	280	31	16
South						
Asia	+ 3.3	+ 8.4	505	490	45	37

During the late eighteenth century, about the time of Adam Smith's *Enquiry into the Wealth of Nations* and Jeremy Bentham's *Theory of Legislation*, radical opinions proposed by the Rational Dissenters advocated that Britain's Poor Laws cushioned the poor and incited idleness [Plumb 1963]. Extreme contradictions between those of social conscience who regard assistance to the poor as a government and civic responsibility, and their adversaries who regard poverty and related destitution as self-inflicted, or as divine retribution for sinful malfeasance, are abundantly evident in today's world. Among nations who enact and enforce benign social policies, the gap between the most and the least affluent is narrower than where successful acquisition of extreme wealth is regarded as just reward for diligent endeavour, irrespective of how the disproportionate wealth is acquired. Among the most acquisitive, Francis Bacon's aphorism that 'knowledge is power' has given way to 'wealth is power'.

Poverty and World Bank Development Reports

Since 1980 at the start of each decade the World Development Report has been devoted to Poverty Alleviation. The 1980 Report was entitled *Poverty and Human Development*, the 1990 Report simply *Poverty*, while *WDR 2000* summarises proposed strategies and actions to reduce poverty, indeed how concepts of human development have changed and become more complex over the past 50 years. During the 1950s and 1960s much emphasis was placed on large investments in physical capital and infrastructures. In the 1970s it was realised that physical capital was insufficient and that investments in health, education and training were essential for people to rise out of poverty. The 1980s witnessed a further shift in emphasis consequent upon the growing debt crisis among poorer nations, together with a worldwide economic recession. Evident was a sharp contrast between progress in the East Asian – Pacific nations on the one hand, and the persistent poverty in South Asia and sub-Saharan Africa. *WDR 1990* called for improved economic management, greater reliance on market forces and investment in health and education.

The *World Development Report 2000* estimates that of the planet's 6 billion inhabitants, 2.8 billion (47 per cent) try to survive on less than $2.00 USD/day, 1.2 billion (20 per cent) on less than $1.00/day. Of these poorest, 44 per cent live in South Asia, 24 per cent in sub-Saharan Africa. Gross inequities between the most affluent and most indigent are reported from many nations. *WDR 2000* estimates the percentage of total national income received by the poorest 20 per cent and 10 per cent, and the richest 20 per cent and 10 per cent in each of the nations listed in Table 9.

The 'Factor [Richest:Poorest]' columns show the per cent of total national income received by the richest 20 per cent and 10 per cent divided, respectively, by the per cent of total received by the poorest 20 per cent and 10 per cent. From these data, the 20 per cent richest in Brazil receive 25.5 times the income of the poorest 20 per cent; the 10 per cent richest in South Africa receive 42 times the average income of the poorest 10 per cent. It is estimated that the total assets of the world's 350 richest people exceed the combined total GNP of the countries that are home to almost one-half of the world's poorest folk. A recent report (2000) by the UN World Institute for Development Economics states that the richest 1.0 per cent of the world's adults owned almost 40 per cent of global assets.

	Poorest		Richest		Factor [Richest:Poorest]	
	20%	10%	20%	10%	20%	10%
USA	5.8	2.3	47.7	32.3	8	14
UK	6.6	2.6	43.0	27.3	6.5	10.5
Canada	7.5	2.8	39.3	23.8	5.2	8.5
Brazil	2.5	0.9	63.8	47.6	25.5	53.0
India	8.1	3.5	46.1	33.5	5.7	9.6
South Africa	2.9	1.1	64.8	45.9	22.3	42.1
Russian Fedn.	4.4	1.7	53.7	38.7	12.2	23.0

WDR 2000 outlines the many dimensions of, and difficulties in quantifying poverty, be it individual or family incomes or estimated levels of consumption. GDP *per caput* in $USD or as Purchasing Power Parity, the quotient of national GDP or GNI divided by the size of population provides little indication of the spread of income across any population.

Poverty in India

WDR 2000 discusses the dimensions and causes of poverty with particular reference to India. The Report describes the desperately poor as people a. with income and assets insufficient to acquire such basic necessities as adequate food, shelter, clothing, health services and education; b. who are voiceless and powerless to communicate with or influence those in authority; c. who possess inadequate material and physiological reserves to withstand sudden adverse shocks: disease epidemics, food shortages, extreme climate change. In India poverty, accompanied by severe deprivation of access to education and health services, is in part attributable to inequities among castes and classes and between the sexes. Data from India's National Sample Surveys (NSS) indicate that despite healthy national economic (GDP) growth, of about 6.1 per cent/year during the 1990s, the pace of poverty reduction across India slowed down. While it is extremely difficult to assemble and analyse demographic and economic data from so large and diverse a population, NSS statistics indicate a widening income disparity particularly between a rising urban 'middle class' and urban and rural poor. *WDR 2000* suggests that comparative assessments of urban and rural poverty across India may be distorted by inadequate research methodologies, and that there may be less difference between rates of rural and urban poverty alleviation than some published data indicate. Nevertheless, though the accuracy of available statistics may be questionable, the inequitable distribution of purchasing power between those who enjoy abundant affluence and those who suffer abject poverty is evident to any frequent visitor to India, Africa and other Asian nations.

Marked disparities in the extent of poverty and between rich and poor is apparent among India's states. Over many years, the State of Kerala, which has provided comparatively superior education and health services, enjoys average life expectancies higher than the District of Washington D.C., despite significantly different average income levels.

ALLEVIATION OF POVERTY

WDR 2000 claims to benefit from earlier accumulated evidence and experience and proposes that poverty be attacked in three ways: promotion of employment opportunity; facilitation of empowerment; enhancement of security. The poorest people are deprived of the most critical and essential of opportunities, namely access to stable remunerative employment. *WDR 2000* lists inadequate access to other opportunities that contribute to rural poverty: credit, roads and communications, reliable electrical power, profitable markets, schools, safe drinking water, hygienic sanitation and health services. In the Bank's judgment, poverty is a complex of many unavailable and inaccessible opportunities, and economic status must be assessed by a mix of quantitative and qualitative criteria. To alleviate poverty, governments must act to reduce inequalities in access to essential opportunities and resources. Sadly, alleviation of poverty by provision of more equitable access is slow to come about, as the rich and powerful exert greater influence over politicians and government policies than do poorer citizens.

Empowerment

WDR 2000 states, as an unspecific generalisation, that empowerment requires participation of all citizens: the poor, middle class, the rich and powerful in formulating decisions that will influence national and individual survival and development. 'Empowerment' is a popular word among people who write about social and economic development, a word of no precise meaning. Literally, empowerment means 'giving power to'. Does the Bank's proposal mean that

a. all citizens actively participate in decision-making and intend that there be significant decentralisation of power, authority and responsibility from national governments to local communities?
b. local communities be granted the resources to pursue and fulfill their decisions?
c. on all major issues there be a formal Swiss-style referendum?
d. all people must rely on their nationally elected or appointed governments to enact and enforce legislation that they, the government officials, decide is in the greatest public interest?

In so many self-styled democracies, rich and poor, 'the greatest public interest' is in practice interpreted as what is in the greatest interest of those who enjoy political power and those who directly support them financially and/or ideologically.

Security

WDR 2000 describes 'security' as 'reducing vulnerability to economic shocks, natural disasters, ill-health, disability and violence ... it requires effective national action to manage the risk of economy-wide shocks and effective mechanisms to reduce the risks faced by poor people ... improving the assets of poor people, diversifying activities and providing insurance mechanisms to cope with adverse shocks.' *WDR 2000* further states that among these there is no hierarchy of importance, all of the elements being complementary. "Helping poor people to cope with shocks and managed risks [enables] them to take advantage of emerging market opportunities".

WDR 2000 advocates a comprehensive approach but admits there is no simple all-embracing blueprint for poverty alleviation; each country must prepare its own policies that reflect national priorities, communal and individual needs, resources available and what actions and investments are practically possible. *WDR 2000* encourages private investment, technological innovation, support of micro-enterprises and small business, access to affordable credit, training in technology and business management, creation and maintenance of essential infrastructures. It recommends that poor nations support agriculture, essential services and expand into international markets. *WDR 2000* reiterates and insists that women in all nations and communities be given more equitable rights and opportunities.

Though there is little in the *WDR 2000* recommendations with which to disagree, they cover a vast spectrum and are stated in broad generalisations. Among 'opportunities' listed, access to stable remunerative employment must rank as a high priority. A statement attributed to a British Indian Army officer in 1856 – "Indian famines are famines of work, not of food. Where there is work there is money for food" – was eloquently elaborated by Mahatma Gandhi in the early 1920s when he wrote: "To a nation starving and idle, the only form in which God dare appear is as work and a promise of food as wages".

The Bank authors mention the need for credit but do not emphasise that those who operate and are employed in rural agribusiness, as producers, processors and/or vendors, need available access to technical, financial and marketing advisory services, services that are permanently accessible in their communities not in some distant government laboratory or delivered in the form of brief, infrequent, transitory visits.

The litany of potential insecurities defies interpretative action. Not even the most affluent, scientifically and technologically advanced nations can predict or insure against all possible sources of natural catastrophe: earthquakes, droughts, tempests; of disease epidemics among humans and farm animals; of civil unrest and military conflict. How can equitable rights and opportunities for women be ensured in a male-dominated world where so few women hold high office in governments, in industry or commerce? *WDR 2000* does not suggest how to change communities where religious and/or social traditions deny women their rights to any semblance of equality or equity with their fathers, husbands and brothers. Nor does *WDR 2000* suggest how the rich and powerful in any nation can be persuaded or forced to be sensitively and practically sympathetic to the needs and opinions of their less fortunate neighbours, to legislate for equitable access to resources and opportunities among all citizens.

NEGLECTED OPPORTUNITIES FOR WOMEN

Among many nations, women are demonstrably more prudent and effective managers of money and business affairs than are men, a fact tragically ignored in many development programmes. Until the nineteenth century 'Economics' in English meant management of the household, the same meaning as the Greek word ïéêïíïìåéá (oikonomeia) from which it is derived. Women, who efficiently managed their households, were the first economists. Throughout most of Africa and much of Asia it is women who manage and control the markets. Some time ago this author proposed to an African finance minister, a man with a PhD. in macro-economics, in order to realise national prosperity, his most constructive action would be to hand over his office to the principal 'market mammy'.

CONSTRAINTS TO TRADE AND HOPES FOR REFORM

Exacerbating poverty among poorer nations, *WDR 2000* estimates that tariffs and subsidies by OECD governments cause annual losses in welfare of over $20 billion USD, equivalent to 40 per cent of all aid in 1999. Past experience is not a cause for optimism that "the developing world and the international community will work together and combine their experience and knowledge to ensure more rapid progress towards ending poverty during the twenty first century". Discussions reported from a recent meeting of the World Trade Organisation offer hope that nations' subsidies to agriculture, a serious impediment to equitable international trade, may gradually be reduced and eventually eliminated. Recognising the political influences exercised by American and European farmers, it is unrealistic to expect any significant reduction in the $350 billion expended annually to subsidise agricultural exports from the United States and the EU.

WDR 2000 ends by stating that concepts of development cooperation are being reformed, new patterns of aid assistance are being devised and tested. The Bank's stated belief in a revived commitment by the international community to alleviate poverty is hardly borne out by present available evidence. During the 1990s among most OECD nations' ODA as a per cent of GNP was in decline.

The world's most powerful and affluent nation invests many billions of dollars annually in armaments and military adventures, less than $12 billion in foreign aid, most of which goes to a nation which by any rational definition cannot be described as low-income, underdeveloped or developing.

According to the International Institute of Strategic Studies, in 2003 total world expenditures in armaments exceeded US$997 billion, a sum more than 17 times the total contributions of OECD nations to aid designed to alleviate chronic poverty and disease. Due to continuing conflicts in Iraq, Afghanistan and elsewhere, investments in armaments will persistently rise while aid directed to reconstruction of Iraq and Afghanistan will overshadow international development as perceived by Pearson, Brandt and Brundtland.

UNDP HUMAN DEVELOPMENT RECOMMENDATIONS

HDR 2003 recommends for people to be lifted out of poverty, governments and international agencies must:

a. invest in health care, education, and ensure access to clean water and hygienic sanitation
b. increase agricultural production and productivity
c. improve rural infrastructures

d. adopt policies to stimulate labour-intensive rural industries
e. ensure equity among all citizens, particularly between the sexes
f. legislate to ensure sustainable environments
g. promote competitive industries to process goods for export.

General Comment

Poverty is both a cause and a result of insecurity: economic, food and health insecurity. Abject chronic poverty with consequent insecurity are fundamental causes of so-called terrorism. When millions of young people are denied access to employment with income sufficient to provide for their basic needs, when their territories are invaded, their property destroyed or taken from them by more powerful nations, they will inevitably resort to violence against their oppressors. 'Terrorism' will persist and become ever more violent so long as powerful nations, ignoring the edicts of the Atlantic Charter, seek to inflict their beliefs and concepts of governance on weaker folk by military force.

The Millennial objectives of reducing poverty by half are admirable but are of questionable realisation. This sorry state will persist so long as those with wealth and power are resistant to more modest and conservative consumption, to sharing their national and personal resources with people less fortunate both within and beyond their national boundaries. As Mahatma Gandhi so sagely observed: "The Planet's resources are sufficient to satisfy everyone's need but not everyone's greed".

The unwillingness of the most affluent to share their possessions with those less fortunate emphasises that redistribution of assets and more equitable access to resources and opportunities must be enacted by governments, encouraged by the people who elect them. In the wise words of the late Dr B.R. Sen, former Director General of FAO, "One person's poverty is every person's poverty; one person's need is every person's need."

7

Sustainable Agriculture

Diverse and Divergent Concepts

Recent years have witnessed an immense outpouring of published literature under the general heading of 'Sustainable Agriculture'. In preparing a report on "Sustainable agriculture and food security" for the 1992 UN Conference on Environment and Development a scientific Commission of the International Council of Scientific Unions (ICSU) reviewed some 300 related publications [ICSU-CASAFA 1991]. The Commission reported how 'sustainable agriculture' is treated from a diversity of scientific, pseudo-scientific, biological, ecological, ideological and philosophical perspectives. The following are a sample of relevant significant publications referred to in the Commission's report: Agriculture Canada [1989], Altieri [1983], Boeringer [1980], Brown [1987], Brundtland [1987b], Bunting [1987], CGIAR/TAC [1989], Edwards et al [1990], FAO [1984], Harwood [1990], Jain [1983], NRC/NAS [1989], NABC [1989], Rodale [1983], Reganold and Papendick [1990], Swaminathan and Sinha [1986]. During the succeeding decade many more have been added.

As defined by the Brundtland panel, a sustainable agricultural system manages essential resources so as to satisfy the needs of all people presently dependent without compromising the needs of future generations. It is when authors try to prescribe agricultural production systems claimed to be of global ecological adaptability that confusions and contradictions arise. Geological and anthropological evidence indicate that, from the time of creation, the earth, its environment and inhabitant creatures have changed, sometimes gradually and evolutionary, sometimes in extreme and disruptive patterns. Many scientists contend that human activities are accelerating environmental and climatic change, some attributable to developments in agricultural technologies. The ICSU-CASAFA report illustrates how the relevant literature presents divergent opinions and recommendations based on experience with different locations, ecologies, technologies, environmental, climatic, social and economic circumstances. Sustainable systems are prescribed by commentators of variable international experience, some seemingly motivated more by ideological persuasions than from a logical, systematic scientific perspective.

Systems and technologies proposed for any farming community should be assessed in relevant context and require a comprehensive analytical understanding of the existing technologies, resources, constraints, prevailing ecological and climatic conditions; farmers' individual and collective capacities to tolerate the inevitable risk of adopting unfamiliar crop types or systems of cultivation. Farmers who produce surplus to their families' subsistence needs must have timely access to market information: the state of supply and demand, quantities and qualities demanded, current competitive prices, seasonal and unanticipated fluctuations in demand.

The ICSU-CASAFA report describes cultivable land as a fast-shrinking resource, and presses the need to raise food crop yields and farm productivity throughout Asia, Africa and Latin America. The report specifies the following urgencies:

a. to protect and enhance the natural resource base: farm land, surface and ground water, genetic diversity;
b. to reclaim and restore fertility in land degraded through misuse;
c. to promote systematic integration of production with post-production activities and to stimulate establishment and maintenance of rural agro-industries;
d. to devote more research to land capacity classification, soil and water management, to ecologically conservative land use.

The report illustrates the diversity, lack of consistency and many contradictions among definitions of, and prescriptions for, sustainable agriculture. It states: "It is unhelpful to propose any particular production system as sustainable in all ecologies, locations and societies. No specific system is suitable and sustainable for all circumstances ... Donor agencies should not seek to impose on poor farmers in developing nations systems designed to remedy wasteful practices in more affluent agricultural economies." The report contends that donors assign greater emphasis to agricultural production technologies than to post-production systems; to policy, social and economic considerations. The report urges far greater support for post-production systems; for technologies, logistics and economics of preservation, processing, transport, distribution and utilisation.

Other Commentaries and Recommendations

A publication commissioned for the 1995 World Food Summit cites various proposed prescriptions and purposes for sustainable agriculture [Hulse 1995]:

a. to generate agricultural products adequate and acceptable in quantity, variety and quality;
b. to maintain environments favourable to humans and other organisms;
c. to prevent pollution of surface and ground water; to protect wild life and animal rights;
d. to prevent despoilment and degradation of fertile land by erosion, urban spread and activities unfavourable to agriculture;
e. to establish and maintain rural infrastructures essential to agricultural production and marketing;
f. to protect natural ecosystems and ensure long-term conservation over short-term exploitation;
g. to promote nutrient recycling and ensure a balance between immediate use and long-term stability.

Some observers, mainly from developing nations, insist that 'sustainable' is not synonymous with 'low-input' systems that reject all chemical inputs. Nutrients derived solely from animal manure and composted biological waste may be insufficient to maintain fertility in poor soils, to support increased crop production needed to feed expanding populations. Modern technologies must serve to complement not replace traditional practices [Jain 1983].

Based upon his extensive international experience, Harwood [1990] presents an impressive historical review of proposed sustainable agricultural systems. He suggests the word 'sustainable' implies a steady state to be realised only by defining long-term goals. He states there are almost as many opinions about sustainable production as there are authors, many opinions being based on too little hard data and little understanding of variability in resources. He suggests many recommendations are simple and devoted to single issues rather than total complex systems. He considers it futile to seek any widely applicable definition of sustainable agriculture and proposes the following as a workable description of 'sustainable': "Agriculture that evolves indefinitely toward greater human utility, more efficient resource use, an environmental balance favourable to both humans and other organisms." Harwood deplores the excessive use of resources and lack of concern for ecologies evident in intensive production systems practised by many American farmers.

Conservation and Preservation

As a general guide to what is sustainable, the following are deserving of particular attention [Swaminathan 2001].

- *Soil health*: physical, chemical, microbiological properties and susceptibility to erosion.
- *Water quality*: water to be used for irrigation must be of low salt concentrations.
- *Plant health:* crops must be protected from insects, other pests, pathogens and competing weeds. Pests can be persistent year round in tropical climates.
- *Genetic homogeneity*: genetically homogeneous crop types cultivated over large areas are susceptible to pests and pathogens that eventually overcome a plant's natural resistances.

- *Abiotic stresses*: soil salinity, flooding, water logging, periodic drought, chemical contamination are serious impediments to crop cultivation.
- *Post-harvest management*: uniformity in maturity, appearance, post-harvest stability during storage and transportation are properties neglected by governments and many donor agencies despite their greater importance to urban consumers and processing industries.

All harvested crops, livestock and fisheries are in varying degrees perishable, critical conditions being active moisture content and ambient temperature. In perishable produce, deterioration starts soon after harvest and in tropical climates progresses rapidly if preservation processes are not applied. As urban communities expand, as spatial and temporal distances between rural producers and urban consumers increase, the probability of post-production degradation increases. Primary processing close to where perishable crops are cultivated serves to reduce avoidable waste and to provide off-farm employment for rural people.

Swaminathan refers to the potentially adverse consequences of climate change, the need for genetic selections that reduce dependence on chemical inputs, research to devise remediation methods to reduce soil and water pollution, for genetic modifications to transfer such useful traits as tolerance to salinity from mangrove plants to cultivated food crops. Tolerance to soil salinity would extend Asian land areas suitable for crop cultivation.

Agricultural Resources

Resources essential to agriculture may be classified as

Internal: Resources that exist and are renewable within a farm's confines;

External: Resources purchased, contracted for, and imported from outside the farm.

Resources can be sub-classed as a. natural, b. synthetic and manufactured, c. socio-economic, d. human [ICSU-CASAFA 1991].

Internal Resources

Land classification and soil characteristics are critical determinants of practicable farming systems, systems influenced by a. land type and profile, whether flat, sloping or undulating, naturally wet or dry, with or without drainage; b. soil properties: structure, depth, density, chemical and organic composition, acidity, water retention and permeability. Efficient land management restricts erosion and structural degradation; maintains fertility and inhibits loss of plant nutrients. Prudent water management restricts pollution by agricultural chemicals and organic effluents, conserves surface, fossil and ground water, prevents overdrafting of rechargeable aquifers.

Internal flora include a. natural vegetation, some useful as forage or fuel; b. cultivated crops; c. soil and air-borne microorganisms, some that symbiotically stimulate plant growth, others that are harmful pathogens.

Internal fauna comprise a. natural wild life: i. productive organisms: insect pollinators, predators and parasitoids that attack crop pests, earthworms that improve soil properties; and ii. destructive organisms: arthropods and rodent pests, some as vectors for pathogens; b. livestock husbanded for food, fibre and on-farm work.

External Resources

Whenever the load on carrying capacity exceeds the internal resources, external supplementary resources become necessary. External resources may be purchased outright or obtained under contact. External resources can be employed to modify land profiles, construct water catchments, plant trees, and provide machines and equipment for irrigation and other functions. External power from electricity or fossil fuels may be needed to drive machines; agricultural chemicals derived from fossil fuels provide fertilisers, costs being influenced by the world price of oil. Use of agricultural chemicals has persistently increased; excessive use can be hazardous to human health, to other organisms and to environmental stability. Bio-fertilisers, bio-pesticides and biological control of pests reduce dependence on agricultural chemicals [ICSU CASAFA 1991].

Agricultural Land and Soil

Soil is a living system: its structure, composition and biological diversity must be understood, protected and prudently managed if productive agriculture is to be sustainable [Brown and Wolf 1984]. The proportion of available arable land already cultivated varies from about 85 per cent across Asia to 22 per cent in sub-Saharan Africa.

However, in Africa, land classed as arable includes large areas of tropical forest, extensive humid ecologies heavily infested with tsetse fly, the vector for the pathogens that cause sleeping sickness in humans and anaemic debilitations in cattle. Tropical forest soils are fragile and rapidly eroded when deprived of tree cover. Effects of soil degradation include debasement of physical structure; erosion by wind and water; depletion of organic matter and natural nutrients; toxicity from salinity, alkalinity, acidity and chemical contaminants. Much soil degradation results from avoidable, human carelessness. [ICSU-CASAFA 1991].

Of the world's total ice-free land, an area of roughly 13.4 billion hectares, the potential arable land is barely 3.2 bn ha [24 per cent of the total] of which about 60 per cent is of low productivity, susceptible to many stresses that inflict degradation. It is estimated every year some 20 miilion ha become uneconomic for farming, and an additional 6m ha are degraded beyond practical reclamation [IBSRAM 1991].

An international assessment of soil degradation reports that more than 1.9 bn ha of world soils have suffered some form of degradation, ranging from light to extreme, most resulting from mismanagement, deforestation, over-grazing, over-exploitation and industrial chemical pollution. Of the Planet's degraded soils, 38 per cent are in Asia, 25 per cent in Africa, 15 per cent in Central and South America, 11 per cent in Europe (including eastern Europe) and 5 per cent in North America [ISRIC 1991].

More than 110m ha of Indian soils are significantly degraded: 81 per cent by wind and water erosion, 13 per cent from loss of fertility, 4 per cent from salinity, 3 per cent from water logging. The area degraded represents about 30 per cent of

India's arable land [Swaminathan 1992]. Of Canada's territorial area of 10 million square km, less than one per cent is high quality arable land. It is this land, best for productive agriculture, located in Canada's most benign climatic zones, that is most favoured for urban and industrial development. The World Resources Institute estimates more than 50 per cent of the world's arable land has been severely degraded [WRI 2004].

Agricultural systems must be sufficiently sustainable to provide for the needs of inexorably expanding populations, with greatest growth in urban communities while rural producers proportionately decline. However defined, by purpose or practice, sustainable agriculture is wholly dependent on fertile arable land and safe water, both disastrously degraded, wasted and misused. Arable land and unpolluted water are fast becoming the main determinants of future sustainable agriculture.

Variations on the Sustainable Theme

Much of the literature on sustainable agriculture originates in affluent industrialised nations and relates to intensive highly mechanised production systems that utilise substantial quantities of synthetic chemical inputs. Observers critical of subsidised intensive systems recognise in their nations less urgency to raise agricultural produce than to restrict disruption and degradation of natural ecosystems and resources. Though 'sustainable agriculture' is widely favoured, the following are other designations with reputedly similar objectives that have emerged in recent years.

Alternative Agriculture

Literally interpreted, alternative agriculture describes any production system that differs from what is commonly practised. In some instances it is used to imply ecologically benign systems less dependent on agro-chemicals, other industrial inputs, more reliant on nutrient cycling, biological nitrogen fixation, propagation of natural enemies of crop pests and pathogens, conservation of soil and water quality, of biodiversity, protection of valuable symbiotic relations and biological interactions among crops, pests and pest predators.

Biodynamic

These favour systems that use compost and humus to enhance soil structure and fertility.

Ecological, Eco-biological, Socio-ecological

These are stated to be based on agricultural ecology harmonised with human ecology, to be protective of environments, to discourage use of chemical fertilisers, herbicides and pesticides.

Low-input

Advocates propose least possible use of agro-chemicals, fossil fuel energy, mechanisation, commercial certified seed and pedigree livestock. They recommend on-farm breeding and seed multiplication, composting of farm waste and crop residues, animal and green manure. Interpreted literally, low-input systems would revert to primitive labour intensive closed farming systems.

Low-input Sustainable [LISA]

Proponents assert that productivity is raised by more efficient use of certified seed, energy, mechanisation and other off-farm inputs; by soil and water conservation, conservative tillage, multiple cropping mixed with livestock, systematic recycling of manure and crop residues.

Others designated 'Regenerative' and 'Permaculture' appear similar to 'Biodynamic' and 'Organic Agriculture'.

Organic Farming

Strictly speaking, all agricultural production is 'organic' in that every product of agriculture is composed of organic substances. Among the earliest proponents of organic farming were two Britons, Sir Albert Howard and Lady Eve Balfour. Before World War II Howard advocated adoption of production systems that used no synthetic fertilisers or pesticides. For more than 30 years, on her research farm, Lady Balfour conducted experiments in which conventional British farming systems were compared with 'organic' systems free from agro-chemicals. Rodale, the American high priest of organic farming, may have received his inspiration from the published work of Howard and Lady Balfour [Rodale 1983].

Food products from organic farming represent a relatively small but slowly growing proportion of foods purchased in Europe and North America. Across the EU, the agricultural area devoted to organic farming has increased from 0.5 per cent in 1993 to *ca* 3 per cent in the year 2000. The value of organic foods sold in the United States is estimated at $10 billion, across the EU at 8 bn Euro. The total value of retailed processed foods in the United States is over $500 billion, in the EU more than 350 billion Euro.

The *Codex Alimentarius* guidelines for organic farming specify the following objectives:

- to enhance biological diversity;
- to increase biological activity and maintain fertility in soils;
- to restore soil nutrients by recycling organic wastes;
- to rely on local renewable rather than imported manufactured resources;
- to promote healthy use of soil, water and air;
- to minimise pollution from agricultural practices.

Despite the interest of environmentally conscious consumers, rational development of organic farming has been hampered by confusion among concepts and definitions, comparable to diverse interpretations about 'sustainable agriculture'. A comprehensive review and state of progress among EU countries has been published [ECDGA 2000]. Another European publication estimates sales of organically produced foods among EU members at about 1 per cent of total food sales. It also comments on the higher prices for organic foods charged in supermarkets. In France price differentials between organic and non-organic produce range from 10 to 40 per cent for dairy products, close to 50 per cent for fresh fruits and vegetables [EEC 2002]. In an 89-page document the European Council presents regulations that specify permitted production, labelling and inspection practices for organic foods produced and sold in EU markets [EEC 1991].

The general principles of organic farming are unquestionably admirable: harmonisation of agriculture with nature, protection of ecologies and environments, conservation of soil fertility and water resources, recycling of wastes, humane husbandry of livestock. At present organic foods are purchased by a small proportion of dedicated consumers who can afford the higher prices. Prices may gradually fall if and when demand and production increase. Whether or not strict adherence to EU regulations, total prohibition of all synthetic inputs and transgenic crops can ensure sustainable food security for all the poorest nations and communities is open to question.

Chemical Fertilisers

While those who claim to practise or encourage organic farming vary in specific recommended practices, most urge systems more environmentally benign, relatively free from synthetic pesticides, growth hormones and antibiotics. Whether chemical fertilisers should be permitted is not consistent among organic farming advocates. Fertilisers provide nitrogen, phosphorus, potassium and other elements to soils naturally deficient, or to replace soil nutrients depleted by growing crops. Chemical fertilisation began with importation of guano from sea birds in Peru and nitrates from deposits in Chile.

In addition to the main constituents – carbon, hydrogen and oxygen – 13 elements are known as essential plant nutrients largely taken up through root systems. Most essential nutrients had their origins in the earth's crust from which soils were formed by weathering when nutrients were released, dissolved in soil moisture, then absorbed by the root systems of wild and cultivated plants. The chemical composition of the earth's crust is highly diverse across the Planet; soils in different locations vary significantly in plant nutrient content (ICSU-CASAFA 1991).

Chemical fertilisers are compounded to provide nutrients in which soils are naturally deficient or depleted by cultivated crops. Nitrogen, essential for all plant growth, originates in the 75,000 tonnes of the gas which float above every hectare of the earth's surface. Elemental atmospheric nitrogen cannot be taken up by plants directly but, by complex biochemical reactions and various organisms, must be converted to ammonia or soluble nitrates. Microorganisms known as Rhizobia live in root nodules of leguminous plants where they convert atmospheric nitrogen into ammonia to fertilise the host plant. In industrial processes, elemental nitrogen is 'fixed' to produce chemical fertilisers in a form that release nitrogen from soil as and when needed by cultivated crops.

Substantial energy is absorbed in conversions of nitrogen into ammonia. In a microbial conversion, energy liberated from fourteen moles of adenine triphosphate [ATP] is required to produce one mole of ammonia. In a Haber-Bosch industrial reactor, pressures of *ca* 20 million Newtons/m^2 at 500°C energise the combination of nitrogen with hydrogen. Some nitrogen-fixing bacteria obtain their energy from organic materials in soils; in legume root nodules Rhizobia derive their energy from the host plant in return for which they supply the plant with accessible nitrogen. As Rhizobia fix nitrogen in legume root nodules, therefore, some organic farmers rotate or intercrop legumes with cereals that lack the physiological and biochemical mechanisms to fix nitrogen. There is hope that biochemical capability to fix atmospheric nitrogen may eventually be transgenically implanted into cereal crops.

Other Sources of Nitrogen

Other nitrogen sources include animal manure; bio-fertilisers such as leaf mulch; compost from crop residues, in some instances assisted by earthworms as agents of conversion; green manure from leguminous plants and trees; such aquatic organisms as azola and blue-green algae. Manure, compost, mulch and other bio-fertilisers provide organic and fibrous materials beneficial to soil structure. Nutrient concentrations vary and rates of release from bio-fertilisers are not easily controlled, being influenced by soil temperature and moisture content. Release of nitrogen is generally slower from animal manure and composted residues high in lignin than from chemical fertilisers.

It is difficult to collect, store, distribute and utilise animal manure without significant nitrogen loss resulting from liquid run-off, leaching by rain, and volatilisation of ammonia. Manure stored in the open air in humid tropics with high rainfall can lose 70 per cent of nitrogen in 24 hours. Manure from cattle driven by pastoralists in semi-arid regions quickly dries and is blown away. Where wood is scarce and kerosene is expensive, animal manure may provide the cheapest source of household fuel. In contrast, chemical fertilisers are stable until used, designed and compounded to release nutrients as specifically needed in desirable composition and concentration.

Nevertheless, excessive application of chemical fertilisers may cause soil degradation, toxicity to earthworms and other useful soil organisms, and by leaching into ground and surface waters may cause insult to human health and harm to fish stocks.

In northern European countries, where high cattle stocking densities are common, serious nitrification of ground water results from manure accumulations that freeze during the winter and release high concentrations of water-soluble ammonia during the Spring thaw. Farmyard manure is a source of various coliform microorganisms some of which are pathogenic. An outbreak of listeriosis in Canada was traced to raw salad vegetables fertilised with farmyard manure.

Contentious Issues

Closed farming systems with crops entirely sustained by recycled nutrients from within the farm perimeter are appealing and deserving of encouragement. FAO estimates that agricultural production must grow by at least 3 per cent annually over the foreseeable future if all the world's people are to be adequately fed. It is doubtful if closed farming systems applied worldwide can come close to realising FAO's objective.

A defect frequently encountered in the 'sustainable' literature is failure to define specifically what or who is to be sustained. Is 'sustainable agriculture' to be interpreted as maintaining a constant level of agricultural productivity, or does it entail production consistently sufficient to sustain an expanding, diversifying population? As shown in Table 10, during the 1990s across the world, and in several regions, population grew faster than cereal grain production [WRI 2004].

1990 to 2000 per cent change

	Cereal production	Average yield	Population	Fertiliser use
World	+9	+15	+16	[-3]
Asia	+19	+13	+19	+35
Africa	+18	+6	+31	[-5]
China	+10	+16	+15	+42
India	+21	+21	+25	+66

In at least ten African nations local cereal production decreased during the 1990s, a matter of serious concern since cereals provide the main source of food energy for the world's poorest people. The International Food Policy Research Institute predicts the demand for cereals across all developing nations will rise by about 50 per cent between 1997 and 2020 [IFPRI 2002].

The IFPRI publication forecasts that between 1997 and 2020 average cereal yields in developing regions will drop from 1.9 to 1.2 per cent/year; demand for meat will increase by more than 90 per cent. Nutritionally, cereals are utilised more efficiently when eaten directly by humans than when fed to animals to produce meat for humans. Unless all livestock raised for meat are fed from natural pastures, food resources would be more conservatively utilised if all humans were vegetarians! IFPRI makes the gloomy prediction that by 2020, both China and India, the world's two largest nations that house almost one-third of all humankind, will become net importers of cereals [IFPRI 2002].

Sustainable Agriculture: Static or Dynamic?

Agriculture cannot be considered sustainable if the quantity and quality of its products are inadequate to sustain the healthy survival of all who are dependent. Sustainable agriculture must be assessed by dynamic not static criteria. Simple maintenance of the *status quo ante* cannot ensure food security for expanding populations. From reliable assessments it appears that more than 5 million pre-school children throughout the world die every year from maladies caused or aggravated by malnutrition

[IFPRI 2002]. Many causes of the melancholy state of, and forecasts for, future food and agriculture are offered:

a. a selfish indisposition of the affluent, both among and within nations, to be more conservative in their consumption habits, to share essential resources more equitably;

b. politicians who, to retain power and influence, invest more in armaments than to sustain more efficient systems of food and agricultural production and distribution;

c. sects more intent on inflicting their particular ideologies on the rest of mankind than in helping to alleviate poverty, social and economic inequities;

d. self-proclaimed authorities who advocate a specific set of production principles or a production system sustainable for the whole world.

Sustainable Agriculture in the United Kingdom

A report submitted to the British Minister for Environment, Food and Rural Affairs by a policy commission on the future of farming and food in Britain begins with a 'Vision' of desirable objectives and conditions for food and agriculture [PCFFF 2002].

a. British farming and food sectors must be profitable, able to compete internationally, be good stewards of the environment, and provide a healthy, nutritious diet accessible to consumers in all income groups.

b. Production, processing and marketing must be holistically integrated, with more effective inter-communication among them.

c. The government is responsible for food safety, nutrition policy, animal welfare and environmental protection.

d. Public policy must recognise the interdependence of town and country.

The Report then compares its idealistic 'Vision' with the 'Present Situation' and states:

a. Farming and food industries [in Britain] are on a path that cannot be sustained over the long term.

b. Because of subsidies and price supports imposed by the EU Common Agricultural Policy, British and EU consumers pay more for food than the rest of the world, inflicting particular hardship on poor consumers.

c. The rural environment has been damaged by years of intensive agricultural production; while some farmers have adopted environmentally friendly practices, many cannot afford to do so.

d. Though the 'Vision' emphasises more holistic integration, significant disconnection exists among farm producers, food processors and retailers; unsatisfactory inter-sector relations and communications are detrimental to overall sustainable efficiency.

e. British consumers are uneasy about the wholesomeness and safety of the foods they purchase.

f. many health problems among Britons result from people eating too much of the wrong kinds of foods.

g. farming has declined in importance as an economic activity, farmers' share of retail prices having fallen. The total food and agribusiness industry contributes about 8 per cent, farming contributes less than 1 per cent to GDP.

h. farming is a highly competitive activity; farmers must develop more sensitive communications with consumers.

i. over 90 per cent of foods are purchased from supermarkets, allowing them undue influence over farmers and consumers.

j. the government must accept responsibility to support more economically efficient, sustainable farming and food systems; the government's role is to facilitate essential change, not to finance stagnation and protection of the status quo.

All of the issues raised in the Vision and Present Situation are discussed in considerable detail throughout the report.

Sustainable Agriculture Initiative [SAI]

A few years ago the SAI was launched by the Unilever and Nestlé corporations, since when some 20 food processing and related agribusiness companies have become participants. Though there is sharing of general information, each participating company pursues its sustainable agriculture research and development independently with particular attention to raw materials essential to its brand products.

The quality of every processed food is critically dependent on the biochemical and biophysical properties of the raw materials from which it is processed. The properties of each raw material result from properties genetically inherited, degree of maturity at harvest, post-harvest conditions of handling, storage, transportation and the processing technologies applied in food factories. The SAI programme seeks to establish and maintain close cooperation between farm producers and processors: to improve and, where necessary, modify farming practices to ensure delivery to processors of raw materials of defined acceptable properties. SAI encourages farming practices conservative of critical resources, protective of land, water, the environment, safety and wholesomeness of all raw materials. Listed among SAI objectives are:

a. to reduce risk of harmful residues such as pesticides, hormones and obnoxious contaminants;
b. to ensure safe on-farm working conditions;
c. to improve and expand farmers' knowledge and skills;
d. to preserve soil fertility, minimise erosion, encourage minimum tillage;
e. to promote farming systems that minimise waste and avoid pollution of surrounding environments;
f. to reduce use of chemical pesticides by adoption of biological and biopesticide control;
g. to use chemical fertilisers conservatively and effectively;
h. to conserve and protect scarce water resources;
i. to minimise energy consumed from non-renewable resources;
j. to explore practical economic alternatives to fossil fuels;
k. to protect biodiversity and wildlife habitats, the health and well-being of all farm livestock;

l. to employ environmentally benign production systems;
m. to raise farm productivity and profitability, to increase crop yields and reduce in-field and post-harvest loss;
n. to devise efficient farming systems based on systematic analysis and simulation modelling of all existing and potential biological, technological, social and economic parameters.

The primary objective of the SAI is to ensure long-term availability of agricultural raw materials essential to their industrial processes, materials that provide more than 2/3 of all raw materials used by these industries. SAI participants' dedication to the above objectives is highly commendable, motivated by the need to restrict extensive degradation of arable land; recognising the competition for land from urban and industrial spread, and the desirability of protecting biodiversity, nature reserves and wild life habitats. SAI seeks to respond to the worldwide shortage of fresh water of which agriculture's consumption is close to 80 per cent of total world supply. SAI recognises growing competition from industries and urban communities, pollution of ground and surface waters by agricultural chemicals, municipal and industrial effluents, acid rain caused by industrial toxic emissions.

Guided by overall 'sustainable' objectives, SAI participants' R and D focuses primarily on agricultural crops and livestock of specific importance to their companies' processes. Unilever's concentration is on tea, palm oil, various vegetables and fisheries; Nestlé gives precedence to milk, coffee, sugar and cocoa; Danone to cereals and fodder crops. Environmental protection and resource conservation are also reflected in SAI participants' factory operations, where they pursue energy and water conservation, reduction and recycling of packaging and waste, elimination of refrigerants and other substances potentially harmful to atmospheric and environmental protection and conservation.

Though much maligned by 'anti-globalisation' activists, many transnational food and agribusiness corporations act more responsibly in conserving critical resources, protecting biodiversity, fragile ecologies and ecosystems, in improving agricultural practices and rural economies than can be said of some governments whose subsidies to farmers, damage to ecosystems and natural habitats in pursuit of oil and timber are destructive of ecologies and biodiversity. Governments and development agencies could benefit by following the SAI pursuit of economic, holistic integration of all logistical activities and processes that constitute a benign integrated food system that begins with sustainable on-farm production and ends with delivery of safe, nutritious foods to consumers.

INTERNATIONAL AGRICULTURAL RESEARCH: THE CGIAR

Genesis and Progress

A history of the foundation and early years of the Consultative Group on International Agricultural Research (CGIAR) was published by the World Bank [Baum 1986]. Other publications provide details of the growth and activities of the CGIAR and the family of International Agricultural Research Centres it supports [ICSU-CASAFA 1991, Hulse 1995, CGIAR 2002]. Annual Reports of the CGIAR and individual IARCs are available from the CGIAR Secretariat at the World Bank in Washington.

All of the CGIAR research centres protest their dedicated commitment to sustainable agricultural systems and practices [CGIAR 1995; 2002]. But it is not always clear whether the primary purpose is for each centre to be permanently sustained or to ensure that national agricultural organisations, the developing country farmers and the populations they serve are to be sustained.

The CGIAR had its genesis in agricultural research sponsored by the Rockefeller and Ford Foundations. In 1943, Rockefeller began a bean and cereal improvement programme in Mexico, which eventually brought forth semi-dwarf high-yielding wheat phenotypes and, in 1963, metamorphosed into the International Wheat and Maize Improvement Centre (CIMMYT). In 1960 the two Foundations established the International Rice Research Institute (IRRI) in the Philippines, and later two centres to study tropical agriculture, one in

Nigeria, and a second in Colombia. The CGIAR began in 1971 as an informal consortium of donor and development agencies with no legal constitution. From the start the World Bank has provided a supporting secretariat. Scientific assessments and recommendations come from the Technical Advisory Committee (TAC), a changing international group of experienced scientists with a secretariat housed at FAO in Rome.

In 1971 eleven members pledged $15 million to support the four existing IARCs. By 1976 membership had grown to 26 donors, who contributed $63 million to the four original and seven new IARCs. By 1992, CGIAR has expanded to a membership of 40 with a budget close to $300 million to maintain a family network of 16 IARCs and related programmes. The *CGIAR Report* for 2002 lists 62 donor members, including the three sponsors: World Bank, FAO and UNDP, and states there are now 16 IARCs that employ more than 8500 scientists and support staff. (A private communication indicates that in 2004 there were 15 centres and programmes, supported by over $400 million from 63 CGIAR members). The *2002 Report* refers to five priorities declared at the 2002 Johannesburg Conference: water, energy, health, agriculture, and biodiversity. Following a brief description of activities in each of the IARCs, the report describes Challenge Programmes in which several IARCS cooperate with emphasis on a. water and food, designed to increase productivity of water for food and livelihoods; b. bio-fortified crops for improved human nutrition: breeding crops with higher than normal contents of such micro-nutrients as vitamin A, iron and zinc. The report also describes a Global Conservation Trust, jointly sponsored by the CGIAR and FAO, to ensure diversity among the most important food crops through an endowment to provide funds sufficient to maintain crop genetic resources held in gene banks around the world. It mentions a report published by the International Peace Research Institute in Oslo that states that between 1989 and 1997 most of the major military and civil conflicts around the world occurred in regions heavily dependent on agriculture [CGIAR 2002].

A working group convened at FAO in 1989 by the CGIAR Technical Advisory Committee, with representatives from IARCs and several donor agencies, concluded that, "whereas agricultural production and farming systems have been well served by the CGIAR family, post-production systems, including breeding for functional properties essential to food processors, have been largely ignored."

CGIAR: Original Purposes and Progress

At the outset CGIAR and TAC agreed that the IARCs' research would be devoted to improvement of crops which provide most of the food for developing nations: crops such as rice, wheat, maize, sorghum, several grain legumes, cassava, potatoes and other root crops. Later IARCs were created for research on livestock health and husbandry, fisheries, forestry, agro-forestry, bananas, and a centre for collection and classification of plant genetic resources and the conservation of biodiversity. Each IARC responsible for one or more food crops maintains an extensive germ-plasm bank of broad genetic diversity. National agricultural research and development programmes can select and receive genotypes and phenotypes of crop species suited to their needs and ecologies. Each centre provides training in the disciplines and activities related to its research programme. The CGIAR and its IARCs have assembled the world's largest, most diverse collection of plant genetic resources, amounting to some 750,000 classified accessions.

The early advances at IRRI and CIMMYT, much publicised as the 'Green Revolution', resulted from Mendelian plant breeding which shifted synthesised biomass from vegetative plant parts into the edible seeds. Plant breeders also developed and propagated rice types that matured more rapidly than farmers' traditional varieties. Rice farmers were thus able to plant and harvest two or three rice crops on the same land every year. During the 1960s and 1970s, rice production across Asia more than doubled; India increased its wheat harvest from 11 million tonnes in 1961 to 70mt in 2002 [IEDRC 2003].

Much early criticism complained that those who benefited most from higher yields were large-scale not small-scale producers with limited resources. The world's history of technological progress illustrates that innovative and novel technologies are first adapted and adopted by individuals and industries with access to resources sufficient to tolerate the inevitable risk of investing in unfamiliar practices. Eventually, the higher yielding crops and related production technologies were accepted by a majority of Asian rice farmers [Chandler 1982; Baum 1986; ICSU-CASAFA 1991; CGIAR 1995; Swaminathan 1999].

Farming Systems Research

During the 1970s IRRI scientists laid the foundation for modern concepts of farming systems research, in a programme that began modestly with a few Filipino farmers but gradually spread to include many hundreds of rice farmers in 12 Asian nations. Western agricultural research traditionally began and continued on an experimental farm until promising results were eventually transmitted to farmers via extension services. In contrast, the IRRI farming systems research methods begin by assessing opportunities for, constraints and risks to increasing production by adaptation of new crop types and production technologies. The methodology starts with an analytical assessment of the biological, technological, physical, financial, economic and ecological resources existent among different farming communities; how and with what resources could more productive technologies be introduced with tolerable risk. The IRRI farming systems methodology has contributed significantly to increased crop production while conserving farmers' ecological and economic resources [IDRC 1981; Chandler 1982; Zandstra *et al* 1981].

Estimated Benefits

Evenson and Gollin [2003] estimate results from what are variously designated 'high-yielding varieties' [HYV] and 'modern varieties' [MV] of food crops grown in developing countries. They compare changes in yield and total production during the 'Early Green Revolution (1961–80)' and the 'Late Green Revolution (1981–2000)' and the calculated influence of MVs mostly developed in CGIAR/IARCs. From 1961 to 1980 MVs accounted for 21 per cent of increased yield and 17 per cent of production increase; from 1981 to 2000, for 50 per cent of yield growth and 40 per cent of production increase. Most notable increases attributable to MVs were in Asia, the Middle East and Latin America. Little benefit from MVs was realised in sub-Saharan Africa where most increases resulted from expanding land under cultivation. Using a mathematical simulation model, the authors estimate that, without MVs in the period 1960 to 2000, food crop yields in developing countries would have fallen. The authors also claim that without MVs food prices would now be significantly higher throughout the world [Evenson & Gollin 2003].

Subcontracting to National Programmes

An initiative that started in the International Centre for Research on Potatoes (CIP), and later pursued by IRRI, subcontracts programmes and projects to national agricultural research institutions in developing countries. The IARCs have trained thousands of scientists from poorer nations many of whom return to their native lands without funds and resources adequate to apply the knowledge and skills they have acquired. Subcontracting IARC programmes and projects to national research bodies serves a. to reduce operational costs, as developing country scientists cost less than their international counterparts; b. to increase resources and strengthen capabilities in national programmes. For agriculture to satisfy the needs of Africans, Asians and Latin Americans, resources and responsibilities should be systematically transferred from CGIAR/IARCs to national programmes and institutions.

GENETIC MODIFICATIONS

Genetic Modifications and Pest Control

All crops cultivated and wild are susceptible to attack by pests that infest, infect, predate or are parasitic. Most damaging are weeds and other competitive plants, arthropods (insects, mites, and spiders), pathogenic bacteria, fungi and viruses. Pests and predators cause insult and injury to vegetative and reproductive organs; pathogens inflict debilitating diseases; competitive and parasitic weeds deprive crops of essential nutrients; birds, rodents and insects wholly or partially destroy crops in the field and/or after harvest. Over countless millennia wild plants have evolved mechanisms and synthesised substances ('natural pesticides') by which to protect themselves from their enemies. Biophysical resistance mechanisms include tissues too tough or fibrous to permit penetration. Defensive biochemical substances are synthesised by growing plants: some toxic, some repellent to particular pests.

Cultivated plants and their pest enemies exist in dynamic relations. Pests mutate to overcome plants' natural resistance mechanisms and to tolerate chemical pesticides. To maintain resistance, plants gradually mutate to modify their effective defensive mechanisms. Being slower than insect and microbial mutations, it is among wild plants that stable evolved resistances are most evident. Cultivated plants can be protected by genetic modifications; by cultural practices such as intercropping where non-susceptible crops act as barriers to pest transmissions between same species; by biological control: propagation and release of organisms discovered to be enemies of particular pests; by chemical pesticides or bio-pesticides repellent or toxic to specific or a broad spectrum of pests. Protection by classical Mendelian breeding involves hybridisation of a susceptible genotype with a type resistant to particular pests. Modern transgenic techniques transfer and stabilise resistant traits from wild, distantly or unrelated organisms into cultivated species.

During the 1980s increases in food grain yields by classical Mendelian breeding were reaching practical limits; harvest yields were levelling off. In cooperation with the International Rice Research Institute, the Rockefeller Foundation developed and supported an international programme to devise techniques by which desirable genes could be transferred between distantly related or unrelated organisms. The Foundation acquired the active cooperation of an international group of experienced geneticists who worked to discover and devise innovative means of transferring genes between sexually incompatible species. The first objective was to enhance inheritable resistances in cultivated rice.

Of the 24 identified species of the *Oryza* genus, only two – *O.sativa* and *O.glabberima* – are cultivated as food crops.

Over countless aeons the 22 wild *Oryza* species have evolved genetically inherited resistances to pests and pathogens to which cultivated species are susceptible. The objective of the RF-IRRI programme was to transfer genes that encoded for particular resistances from wild to cultivated *Oryza* species and thereby reduce rice farmers' dependence on chemical pesticides. Recognising that many persuasive activists suffer from extreme chemophobia, believing that any substance of biological origin is inherently less obnoxious and potentially harmful to human health than chemical synthetics, the RF/IRRI programme's objective of reducing reliance on chemical pesticides had logical reason.

Unfortunately, certain transnational agrochemical companies adapted the novel transgenic techniques to increase cultivated food crops' tolerance to chemical herbicides, thus permitting affluent farmers to apply higher than normal doses of herbicide to control competitive weeds. These commercial transgenic modifications were thus designed not to reduce but to increase demand for chemical pesticides.

Two international committees composed of experienced biological scientists pointed out that when a particular resistance is transgenically introduced into a cultivated plant, the plant's biochemistry is changed: it synthesises substances novel and foreign to its constitution, substances toxic or repellent to one or more pests. The target crop of the RF/IRRI programme was cultivated rice, the cereal that provides over 70 per cent of food energy for Asians. Much of the rice grown by small-scale Asian farmers is grown, harvested, distributed and eaten without scrutiny by a food regulatory agency.

Some years earlier, potato breeders in Britain and the United States had crossed standard cultivated varieties with Peruvian wild species to transfer resistance to certain pernicious pests. Fortunately, in both instances by happenstance, before the new hybrids were released to farmers and consumers, visiting scientists discovered that the resistances came from excessive and unsafe concentrations of a neurotoxic alkaloid. The new resistant types were immediately withdrawn and destroyed. With this experience in mind the two committees and a working group convened by the Canadian Food and Drug Directorate recommended to the rice breeders that, before release of any novel transgenic hybrids, biochemical analysts be engaged to determine the composition of the substances synthesised to provide pest resistance, to discover whether they were deposited in the edible seeds, or in vegetative plant parts. If in the edible seeds, toxicological and nutritional studies were essential to determine if concentrations present could result in insult or injury to the health of consumers. Had this advice not been totally ignored, subsequent opposition to GM crops might have been less clamorous, persuasive and restrictive of benefits that prudent genetic modifications could bring about [ICSU-CASAFA 1991, H & W Canada 1989, Hulse 1995].

Adoption of GM Crops

A recent report by Clive James, a Canadian plant pathologist, states that in 2005, 54 million hectares (Mha) were planted to GM soybeans, 21.0 Mha to GM maize, 9.8 Mha to GM cotton and 4.6 Mha to GM rape-canola. GM crops are cultivated by roughly 8.5 million famers in 21 countries. The largest GM cultivators, in sequence of quantities produced, are the United States, Argentina, Brazil, Canada and China.

Disturbingly, James reports that 71 per cent of 90 Mha under GM crops are planted to herbicide-resistant genotypes of soybeans, maize, rape-canola and cotton. Though the original intention of the IRRI-Rockefeller GM programme was to reduce dependence on chemical pesticides, in fact the principal beneficiaries are herbicide manufacturers who also produce the herbicide-resistant seeds. The main active ingredient in the most extensively used herbicide is the isopropyl amine salt of glyphosphate. The potential toxicities of the active and other ingredients are still under study. In North America, the permitted residues of pesticides are scientifically prescribed and systematically monitored, but the same cannot be said of many poorer nations.

Assessing the Safety of Genetically Modified Crops

Persuasive publicity opposed to all genetic modifications, particularly among Europeans, whose survival is not threatened by food shortages, could prove detrimental to future agricultural research and to realisation of food security among poorer nations. While admitting that it is easier to discover acute or chronic toxicity than to demonstrate that a food or process can be assured as absolutely safe for all people under all circumstances, because of stringent food regulations, people resident among affluent nations have access to the safest food supply the world has ever known. But what may be demonstrated as 'safe' by reliable scientific analysis may not coincide with what is perceived to be safe by consumers influenced by strident propagandists.

The opportunities for crop improvement by molecular breeding and transgenic techniques are deserving of responsible scientific exploration. During the 1990s, it is estimated that more than 600 different chemical pesticides, valued at roughly $22 billion per year, were sold around the world [Dekker 1991]. Among nations of the EU, North America and Oceania, agricultural use of crop pesticides is rigorously controlled, while residues after harvest limited by law are systematically monitored. Among poorer nations, pesticide use is less stringently controlled and many instances of pesticide intoxications suffered by farmers and consumers have been reported. Broad-spectrum pesticides destroy not only targeted species but also many harmless and useful organisms. To induce inheritable resistances and thereby reduce use of dangerous pesticides is rational and sensible. Several scientifically reliable protocols to ensure safety of GM crops and derived foods have been published [Doyle and Persley 1996; IFBC 1990; IFST 1996; IFT 2000; IFT 2001; WHO 1991].

Natural Pesticides

It has been estimated that some 5,000 'natural pesticides' – substances synthesised by plants to protect themselves – are present in foods sold in the United States. Though many have been identified, it is claimed that too little attention has been given to possible chronic toxicity, carcinogenicity, mutagenicity (disruption of DNA), and teratogenicity (induced congenital malformations) of many 'natural pesticides' [Ames 1983; Ames & Gold 1990]. An objective of many food plant breeding programmes was to eliminate or significantly reduce the concentrations of identified 'natural pesticides' known to be toxic, such as solanin the neurotoxin found in wild potato species. Cultivation of food crops deprived of their natural protection, require larger applications of chemical pesticides.

GM Crops: Potential Benefits

A comprehensive review of opportunities for control of insect pests by molecular biology and transgenic modification concludes that these techniques will lead to improved pest management, facilitate precise studies of insect populations, genetic mapping, insect and crop resistance mechanisms. In addition to creating induced resistance in cultivated crops, genetic modifications of insects offer future means of biological pest control [Evans 1993].

Other potential benefits that genetic modifications of crops could provide include:

a. Greater tolerance to abiotic stresses – Tolerance to salinity can be transferred from mangrove plants into rice and other food crops [Swaminathan 2001]. Rice tolerant to soil salinity could substantially extend the land area across Asia for rice cultivation.

b. Enhanced nutritional quality – More desirable proportions of essential amino acids in cereal and legume proteins; higher contents of micro-nutrients deficient in many poor nations' diets [Swaminathan 2000, 2001].

c. Functional properties to expand suitability for food and other industrial processing – Scientists at the Plant Biotechnology Institute in Saskatoon, Canada, have modified the biochemical and biophysical properties of cereal starches to diversify their industrial applicability [Davis *et al* 2003; Demek *et al* 2000].

These are but a few of the possibilities for agronomic, nutritional and technological modification of food crops that could increase productivity and diversify utility of food crops for the benefit of poorer nations where, without significant improvement to protect and expand land suitable for cultivation, to raise harvest yields, and to reduce losses to pests and other degradations and debilitations, food security for many future millions in Asia, Africa and Latin America cannot be assured.

Closing Comment

Practices and purposes deemed essential for 'sustainable agriculture' differ markedly among practitioners and commentators. As Harwood so cogently observed, there are as many prescriptions as there are authors. Commentators who live in affluent societies should be wary of prescribing remedies for poor folk who exist in fragile ecologies. The need to conserve biodiversity, arable land and water resources is undeniable. But no known crop or livestock production system is sustainable in all places and for all people. The IRRI farming systems methodology of acquiring a comprehensive understanding of what exists before offering remedies for improvement is a worthy example for all to follow.

Notwithstanding the many potential benefits to farmers and to food and health security offered by scrupulously designed and developed GM organisms, there is justifiable concern that the principal beneficiaries of agricultural GM are the producers of herbicides and herbicide-resistant food crops. Instead of reducing chemical pesticide use, herbicide applications are increasing.

In addition, it has been reported that some seed companies are experimenting with Genetic Use Restriction Technology (GURT), also known as 'Terminator' biotechnology. The plants are genetically modified to produce a sterile seed, so that farmers cannot save some of their harvested seed for later planting, but every year must buy new seed from the supplier. This practice is condemned under the Convention on Biodiversity formulated at the Rio conference, but there is evidence that some avaricious companies and the politicians they support and influence have hopes of ignoring the Convention.

SUSTAINABLE FOOD SECURITY

Diverse Definitions

As with 'sustainable development' there is no precise, universally accepted definition of 'food security'. In broad principle a state of food security exists where all individuals, families and communities enjoy consistent access to foods that in quantity, quality and biochemical composition provide hygienic, nutritional adequacy. The Brundtland Panel states: "Food security requires secure ownership of, or access to, food resources and income earning activities, including reserves and assets to offset risks, to ease shocks and meet all contingencies". This may be interpreted as all people having access to stocks and flows of food and cash sufficient to satisfy their basic nutritional needs. Everyone in the world need not be a subsistence farmer, but those who do not produce foods in quantities sufficient to satisfy their families' needs, must possess the means to acquire foods nutritionally adequate, readily accessible and affordable [Brundtland 1987b; ICSU-CASAFA 1991].

The report of the 1992 FAO/WHO international conference on nutrition states: "Food security is defined in its most basic form as access by all people at all times to the food needed for a healthy life.

Food security has three dimensions: assurance of a safe, nutritionally adequate food supply at national and household levels; a reasonable degree of stability in the supply of food throughout each year and from one year to the next; every household having physical, social and economic access to food sufficient to meet its needs."

While there is agreement in broad principle on requirements for food security, concepts diverge significantly in semasiological and biological specifics, they change over time, and differ among specialists and agencies as to what constitutes a nutritionally adequate diet. Non-scientific and quasi-scientific publications frequently fail to make clear distinction among 'hunger', 'malnutrition', 'famine' and 'starvation'; between 'chronic' and 'transitory hunger'. The noun 'security' derives from the Latin '*securus*' which means 'safe', 'free from worry and care'. People blessed with food security are safe from the risk of famine, starvation, chronic hunger and malnutrition. 'Famine' describes extreme food scarcity among large populations. 'Starvation' is derived from an Old English word meaning to die a slow lingering death from cold, disease, insufficient food and/or water. 'Malnutrition' defines a significant insufficiency or imbalance in essential nutrient intake [Latin *'nutrire'* = 'to nurse, suckle or feed'], 'sub-nutrition' sometimes describes people who have too little food. 'Hunger' describes distressing discomfort caused by want of food. 'Transitory hunger' is experienced from time to time by active persons who enjoy healthy appetites. 'Chronic hunger' describes a painful and debilitating condition caused by long deprivation of food sufficient to sustain healthy growth and activity [Hulse 1995]. The 1960s Freedom From Hunger Campaign, the 1974 World Congress' hope of eradicating hunger by 1984, and the FAO Millennium objective of reducing incidence of hunger by 50 per cent between 1990 and 2015, all relate to alleviation of chronic hunger and malnutrition.

Roughly 18 per cent of the world's population (800 million, including 200m children), 25 per cent of South Asians (300 million people) and 35 per cent of sub-Saharan Africans (190 million) are classified by UN agencies as undernourished: their caloric intake is lower than is required for healthy activity. In 1980 30 per cent of all people in LICs were considered undernourished. By the year 2000 this proportion had fallen to 18 per cent [FAO 2000].

Food Security, Income and Poverty

In simplest terms, food security depends upon a dynamic balance among disposable income, demand, supply and distribution. Production of food crops, livestock and fisheries must progressively increase to satisfy expanding and diversifying needs and demands of growing populations. Efficient post-production systems, to ensure safe and effective protection, preservation, transformation and distribution must be economically integrated with production systems. During the early 1960s, affluent OECD nations were home to 34 per cent of world population and controlled close to 90 per cent of global GNP. By 1990, the affluent minority had fallen to 16 per cent of the total but controlled more than 82 per cent of global GNP. Between 1960 and 1990 a new group of middle-income newly industrialised countries emerged, most in East Asia and Latin America, a change with considerable implications for the Planet's economy and future patterns of food demand. In 1990 Asians represented 52 per cent, in 2002 some 59 per cent of world population. Among Asian nations, middle-income communities are growing most rapidly, their diets diversifying with dilating demand for livestock products.

Food insecurity, chronic hunger and under-nutrition are both dismal consequences and clear indicators of extreme poverty. People with money, power and control over negotiable assets rarely suffer from chronic hunger. Government politicians and military generals who suffer from chronic hunger are not a common spectacle around the world. Alleviation of food insecurity requires greater opportunities for paid employment and more equitable access to critical resources and assets among and within nations.

Changing and Conflicting Concepts

A respected observer writes: "In the 1950s ... it was assumed that adequate agricultural production would assure access to adequate food in the market and in the household. In the 1970s it became clear that production and availability alone do not lead to food security, since those who lack purchasing power (from permanent employment income) lack access to a balanced diet. Now it is evident that where availability and access are satisfactory, necessary biological absorption of food requires clean drinking water, environmental hygiene, primary health care and education" [Swaminathan 2001].

Based on these considerations, the M. S. Swaminathan Research Foundation and the UN World Food Programme have published a *Food Insecurity Atlas for Rural India* [Vepa and Bhavani 2001]. The atlas illustrates marked variability among Indian States in food availability, absorption of nutrients, vulnerability to chronic hunger and in sustainable agricultural production. The atlas describes such non-food factors as income earning opportunities, health care facilities, education, sanitation and environmental hygiene.

Due to climatic variability, years with too little or too much rain, damage and destruction by pests, parasites and pathogens, crop and livestock production inevitably change from year to year. To compensate for years of poor harvests governments and/or regional organisations should maintain reserve stocks to carry over surpluses from regions and seasons of abundance to those of scarcity. The cost of holding reserve stocks should be calculated as an insurance investment not an unwarranted expense.

Food security need not be synonymous with food self-sufficiency, a nation's ability to produce all food required by its population. For diverse reasons: climate, topography, soil and climatic conditions, many nations cannot be self-sufficient and must depend on food aid or imports.

Substantial reductions in food spoilage after harvest could be realised by establishing primary food preservation and protection facilities in rural areas where crops are cultivated and animals husbanded. Food saved from spoilage increases total food available and enhances food availability and security.

Public Policy and Food Security

Estimates of 'Global food security', based on the quotient of total possible food production divided by world population, may be a diverting statistical exercise, but is of little practical value. Equitable access to the planet's resources, though highly desirable, never has been nor is ever likely to be a realisable objective. To state that over the next half-century world food production must exceed all that was produced since settled farming began 10,000 years ago is interesting but offers little practical guidance for long-term planning by national governments. Food security can only be assessed and assured for discrete nations, communities and families within communities, a process that is the responsibility of national and local governments. International agencies can help weak and inexperienced governments by defining criteria to be considered and by providing long-term development assistance.

During the 1930s, the League of Nations Committee on Nutrition reported: "In the past, movement towards better nutrition has been largely the result of unconscious, instinctive groping for a better, more abundant life. Now needed is planning for better nutrition policies based on (a) consumption: bringing essential food within reach of all people; (b) adequate supply: production and equitable distribution of all essential food sources" [League of Nations 1936–1937]. The report urges (a) recognition of nutrition policy as of primary national importance; (b) better, widespread education on human nutrition; (c) more equitable distribution of income: "it is the poorest people who are most nutritionally deprived".

A committee of the US National Research Council appears to agree: "An important cause of malnutrition is absence of policies and programmes that foster the best use of available food supplies ... Governments continually make decisions that affect nutritional status with little knowledge of nutritional consequences. Alleviation of hunger and malnutrition depend upon increasing the right kinds of food, reducing poverty and improving the stability of food supplies" [NRC/NAS 1977].

Governmental Responsibilities

Publications that discuss food security are many, vast and voluminous. They suggest primary causes, and propose various courses of remedial action. The League of Nations reports are among many that insist that food security be of highest priority in all governmental policies. Reutlinger [1987] states that food insecurity results from inappropriate macro-economic policies, from local economic and political structures that restrict the ability of households and individuals to have access to foods sufficient to satisfy their needs. Food insecurity may be transitory, lasting for relatively short periods, or chronic and of long duration, a consequence of poor planning to ensure resources adequate to produce or purchase sufficient food. Transitory hunger frequently results from unstable food production systems, sharp rises in food prices, or depression of disposable family incomes. If extreme and extended, transitory hunger can exacerbate into chronic hunger and eventually into famine. Famines have ravaged even when food stocks were plentiful: the 1943 famine in Bengal was precipitated by the price of a limited food supply rising to levels unaffordable by poor people.

Sen [1987] relates food security to entitlements and ownerships; families and individuals must have adequate access and purchasing power. Food security depends not on estimates of national aggregate demand but on fair exchange in international trade of goods and services. The poor must inevitably rely on food aid provided by social programmes. Chronic hunger and malnutrition are less influenced by a nation's total food supply than the degree of equity with which available resources are distributed and made accessible. Food and agricultural policies to ensure food security for all citizens should be a high priority for every government. Malnourished people, too weak to work effectively, eventually impose a financial burden on the nation's economy [Sen 1987].

Strategic Food Stocks

Though some observers argue that food stocks of grain are of value only to alleviate temporary food insecurity, the prime purpose of strategic food stocks is to provide emergency reserves against regional, national, local and seasonal shortages, to stabilise food prices and counteract fluctuations in supply and availability. Reserve food stocks are customarily controlled by government agencies, authorised to maintain quantities sufficient for defined time periods. Reserve stocks must be replenished when they fall below a prescribed safe minimum. Reserve stocks, essential in regions susceptible to frequent drought, provide short-term relief but cannot be considered a panacea for chronic food insecurity.

Santorum and Gray [1993] list factors to be weighed in determining minimum reserve stocks. Stocks stored on farms are generally sufficient only to satisfy farmers' needs from one harvest to the next. Countries within a region that cannot individually afford to maintain national reserves can collectively contribute to and draw upon strategically dispersed food stocks. During the 1980s the Southern African Development Community (SADC) created and maintained a regional food security programme with grain stocks upon which member nations could draw in time of need.

Food Aid

Where many are too poor to buy adequate food, where climatic change, civil strife and other adversities inflict serious disruptions on food production and distribution, food aid may be essential to survival and avoidance of famine. The UN World Food Programme [WFP] is the most efficient distributor of food aid, designed to provide relief or to support development. Food aid for relief alleviates acute stress among people whose food supplies are disrupted by war, civil unrest, enforced migration, severe climatic catastrophes, and/or massive crop failure. Food aid for development provides food as wages for economically productive work and to support basic social services. Food aid for development may support labour-intensive agricultural production, construction of roads, rural infrastructures, amenities and services essential to developing economies. Food aid helps to stabilise food prices, to provide for pregnant women, nursing mothers, school feeding and pre-school infants.

WFP administers food aid most efficiently. Misdirected, ill-conceived bilateral food aid can do more harm than good. One could cite a weary litany of bilateral food aid where indiscriminate dumping of donors' surplus crops and livestock products have depressed prices and discouraged agricultural production in recipient countries. Food aid administered by many bilateral agencies is motivated more by the donor's political and commercial interests than relieving acute or chronic distress among malnourished people.

WFP resources have been seriously overstretched, because of exceptional demands resulting from military, civil and inter-ethnic strife and conflict, mass migrations such as in Darfur, cataclysmic disruptions as produced by the Tsunami. As extreme demands for food aid will inevitably persist, more generous and sustained support for the World Food Programme should be a high priority for all donor agencies.

Nutritional Adequacy

What constitutes nutritional adequacy has attracted scientific and unscientific interest for many centuries. Alchemists searched for but never discovered the *Elixir vitae*. In Britain during the mid-nineteenth century desirable dietary standards were described as: "the least costly foods to prevent starvation and disease during massive unemployment". The League of Nations standards were intended "to marry health and agriculture" [League of Nations 1936–1937]. Our most primitive ancestors, by trial and error, identified certain foods as pleasant and digestible, others as obnoxious, insulting or injurious to health, foods acceptable and foods unacceptable. Our early antecedents chose their foods from plants and animals indigenous to their ecologies and climates resulting in widely different food preferences among nations and communities.

Dietary Standards and Recommendations

Estimates of human nutrient requirements were first codified during the latter half of the nineteenth century. From the time of their establishment, FAO and WHO have sponsored many meetings, committees and working groups to study and recommend desirable nutrient intakes, each meeting re-estimating recommended levels based on new knowledge. At least 50 national, regional and international agencies recommend what they consider desirable intakes of essential nutrients. Recommendations are customarily cited as specific amounts of each essential nutrient to be ingested daily. Some authorities name them Recommended Daily Allowances [RDAs], others Recommended Daily or Dietary Intakes [RDIs], others Recommended Nutrient Intakes [RNIs]. The evolution of recommended dietary standards was reviewed by several authors [Beaton 1991; Hulse 1995; IUNS/CABI 1983]. The purpose of RDAs is to encourage healthy eating habits and to prevent diseases attributable to nutrient deficiencies. Earliest motivations for government dietary standards arose in times of severe stress and distress; to feed nations at war or when suffering economic depressions, widespread poverty and malnutrition.

Recommended Daily Allowances [RDAs] the preferred American designation, are intended to serve several purposes: (a) to assess adequacy of food intake among selected population groups; (b) for national food and agricultural policy formulation; (c) to balance food nutritional quality with human need; (d) to provide adequate food during emergencies, particularly for the poor and disadvantaged; (e) to prescribe and enact standards for nutrient labelling of packaged foods. RDAs were first designed to meet the needs of the majority, often embodying margins of safety to allow for variable needs among people of different ages, activities and conditions of health. It is now customary to prescribe specific RDAs for citizens of different sexes, ages and levels of activity.

RDAs were first calculated by estimating the average physiological requirements for each absorbed nutrient, a value then modified to compensate for incomplete absorption and utilisation, variability among persons, bioavailability among different food sources. RDAs are based on foods consumed in normal diets, not on pure substances or concentrates, and take account of nutrient variability within and among particular foods. At high levels of intake some nutrients, such as vitamin A, may be toxic so that upper limits of safe intake are prescribed. It is difficult to prescribe precise RDAs for all conditions of men, women and children and their variable activities. Experiments on human beings are expensive, time consuming and only small groups of volunteer subjects can be studied in confined controlled conditions. Using humans as experimental animals also raises ethical objections. RDAs are intended for people as they go about their normal daily activities but food intakes and nutrient balance are difficult to determine where subjects move freely and are not confined in a metabolic research facility. Data based on *ex post facto* recall by subjects going about their normal lives are at best approximate.

Despite a vast accumulation of nutritional data, with more sensitive analytical tools available, nutrition remains a relatively imprecise and speculative science. Concepts change as more evidence comes to light. Thirty years ago it was believed military men working in Arctic climates needed more food than those working in temperate zones so Canadian troops in the Arctic north were given packaged rations that provided 5,000 kg Cals/day.

It was later realised that with insulated clothing and the need to move slowly to avoid heavy perspiration that could freeze, caloric requirements were no higher in cold than in temperate climates. Since caloric intake must balance energy expenditure, people who perform heavy work or strenuous exercise need higher caloric intakes than sedentary persons.

People who change from energy-intensive activities to a sedentary life-style tend to eat more than they need. Overeating with resultant obesity is disturbingly evident among the richer nations and affluent persons in less developed economies. It is reported that between 25 and 30 per cent of Europeans and North Americans are excessively obese, at risk of cardiovascular diseases, diabetes and other related disabilities. The obese among the affluent reflect Nerissa's comment in Shakespeare's *Merchant of Venice*: "They are as sick who surfeit with too much as those who starve with nothing."

Nutritional Fashions

Responsible nutritional scientists seek to encourage rational eating habits based on the best available knowledge. The alchemist's pursuit of the *Elixir vitae* is far from moribund among publicists who proclaim all manner of diets guaranteed to ensure good health and longevity [Hulse 1995]. The penchant for nutritional panaceas has been evident for the past 50 years. During the 1950s vitamins were in favour, in the 1960s proteins and amino acids were in fashion, from the 1970s cholesterol and cardiovascular dysfunctions obsessed public attention, next came the health benefits promised by dietary fibre. The present passion is for nutriceuticals, proclaimed in principle some 4,000 years ago by the Chinese, and for 'functional foods' (which beg the question: what are non-functional foods?), neither of which are specifically defined.

Variability in Composition

The nutrient content of every food crop varies among genotypes and conditions of cultivation. Among the world's wheat collection on a dry weight basis, protein contents range from 6 to 23 per cent, lysine (as per cent of protein) from 2 per cent to 4.3 per cent [Hulse and Laing 1974]. A review of over 1,300 sorghum analyses reported protein contents (N × 5.7 DWB) to range from 7.1 to 14.2 per cent, lysine (mg/g nitrogen) from 71 to 212, iron (mg/100g) from 0.9 to 20.0, niacin [mg/100g] from 2.9 to 6.4. Significant variability was reported among samples of pearl millet (*Pennisetum typhoides*), chickpea (*Cicer arietinum*), faba bean (*Vicia faba*), and lentils (*Lens esculenta*)

[Hulse 1991a, 1994, 1995; Hulse and Laing 1974; Hulse *et al* 1980].

Accurate chemical and biochemical analyses are difficult and demand professional competence and experience. Inexperienced technicians are unaware of many sources of error among standard analytical methods. Some 16 sources of error were identified in the classical Kjeldahl method of determining protein nitrogen [Williams 1974].

Changing Recommendations

Since the nineteenth century, recommended daily intakes of all known nutrients have changed among and within responsible agencies. In 1985 a WHO committee recommended a daily protein intake of 0.6g per kg of body weight; a US NRC committee advocated 0.75 g per kg of body weight. Table 11 illustrates how US RDAs have changed since first promulgated in the early 1940s.

US NRC Recommended Daily Allowances

	1945	1989
Protein (g/day)	100	59
Ascorbic acid (mg/day)	100	60
Iron (mg/day)	15	12

UK Recommended Daily Nutrient Intake

	1950	1979	1993
Vitamin A (RE* mg)	1500	750	800
Ascorbic acid (mg)	20	30	60
Niacin (mg	12	15	18
Iron (mg)	12	10	14

* RE = retinol equivalent

A comparison of daily intakes recommended by 39 governments and three international agencies, updated in 1995, illustrates many inconsistencies. Table 13 lists intakes prescribed for young active males (M), for teen-age women (F). Disparities of similar magnitude were evident for all other ages and levels of activity.

Recommended nutrient intakes (RNIs) from 41 countries and agencies

		Highest RNIs	Lowest RNIs
Energy (kcal)	M	3200	2500
	F	3000	2030
Protein (g)	M	105	37
	F	96	31
Vitamin A (mg RE)	M	1500	600
	F	1500	450
Vitamin D (mg)	M	10	2.5
	F	10	2.5
Ascorbic acid (mg)	M	95	30
	F	85	25
Iron (mg)	M	24	5
	F	35	12

Many more people would be classed as malnourished if the highest rather than the lowest or some level in between were the international standard.

For international comparative purposes it is sensible to use the intakes prescribed by WHO and FAO. In light of these many differences and uncertainties, all RDAs, RDIs and RNIs should be regarded as indicative but not scientifically absolute. Disparities among nations may reflect different life styles and activities or various differences of opinion. A Canadian publication states:

"The primary use of nutrient recommendations is to form a basis from which to advise the public on how to choose a healthy diet. It would be a mistake to conclude that those who fail to adhere to each recommendation are doomed to suffer associated chronic deficiency disease, or that by strict adherence consumers will avoid deficiency diseases" [H and W Canada 1990].

Recommended Daily Allowances (RDAs)

RDAs, RDIs and RNIs can be useful as general guides for cognisant officials responsible for formulating national health, food and agricultural policies. For most consumers, RDAs have little practical meaning. Consumers are better advised by illustrated guides to healthy eating based upon mixed diets of foods that are familiar, acceptable and seasonably available [H and W Canada 1992].

Human nutritional requirements, criteria used to arrive at recommended levels of intake, consequences of serious deficiencies, food intolerances and allergic responses are discussed in greater detail in other publications [Beaton 1991, IUNS/CABI 1983, Hulse 1995].

Urban Growth and Food Security

Predictions of the size to which the world population will grow, and for how many persons the planet can provide sustainable food security 50 or 100 years from now, though intellectually diverting, are of little practical utility. One can estimate with reasonable certainty what could be produced with present resources. What may be feasible 50 or 100 years from now is indeterminable.

Of more pressing urgency are the consequences of rapidly changing demographics and rising disposable incomes among urban communities. More than half the people of Asia and Latin America live in large towns and cities, expanding in size and populations all over the world.

Though some food can be raised in city gardens and horticultural allotments, urban communities rely on food safely transported from rural farms, inland and maritime fisheries.

As people migrate from rural to urban communities and disposable incomes rise, eating habits change. Growing demand for processed convenience foods is illustrated by the replacement of cereals, pulses and cassava, hand ground and cooked into some form of porridge, with processed foods such as bread, biscuits, cakes and pasta. Of critical consequence for natural resource conservation is the growing demand for livestock products, which across SE Asia is rising by over 20 per cent a year.

Few of the world's cities were planned and laid out in expectation of the volume and density of vehicular traffic destined to grow progressively worse in all cities on all continents. To alleviate traffic congestion, the City of London Authority has imposed heavy tolls on private vehicles that enter the city on working days.

Too little investment is devoted to improving and maintaining rural infrastructures, to rural-to-urban and intra-urban transportation systems. The longer time taken to transport perishable food materials from rural producers to urban processors and markets, the greater is the probability of biological and economic loss.

Urban food security demands consistent access to a safe adequate supply of acceptable, affordable foods dependent on a. technologically efficient systems of preservation and processing, primary preservation being applied soon after harvest; b. logistically efficient and economic means of storage and transportation; c. reliable markets which retail foods that are safe, of acceptable quality, in unit quantities and prices that all segments of urban society can afford; d. national food and agricultural policies that provide and maintain the infrastructures, services, transportation facilities, social and economic conditions essential to efficient holistically integrated food systems.

Final Observation

The Brundtland principle of providing adequately for present need without detriment to the needs of future generations has particular relevance for food security.

The future need, globally and among specific communities, and future agricultural capacity cannot be predicted, it is therefore, incumbent on all peoples, their governments, industries and institutions, to ensure that resources critical and essential to food production, preservation and distribution be protected and used conservatively. Degradation of arable land suitable for crop production must be arrested, where possible degraded land must be reclaimed. Water, the most critical and scarce of all essential resources, must be protected from pollution and misuse, whenever feasible purified and recycled.

In expectation of unrelenting inter-ethnic conflict, massive migrations, increasing severity of damage by hurricanes and cyclones, together with severe droughts, all consequent upon climate change, the need and demand for food aid is more likely to grow than contract. More generous and sustained support for the World Food Programme must be accepted as an obligation by all affluent nations and donor agencies.

Assurance of sustainable food security for expanding congested urban populations demands efficient integration of all activities and procedures between rural producers, food processors, urban markets and consumers. Urban food security in all nations calls for systematic planning by professionally qualified men and women of imaginative vision. For, as the *Book of Proverbs* so wisely observes: "Without vision the people will perish".

8

Development of Industrial Biotechnologies

DEFINITION AND EVOLUTION

Biotechnologies are defined as 'Processes that produce, preserve and/or transform biological materials of animal, vegetable, microbial or viral origin into products of industrial, commercial, economic, social and/or hygienic utility and value'. Biotechnologists and bioengineers are men and women qualified to design, develop, operate, maintain and control biotechnological processes.

The name 'biotechnology' first appeared in 1897 in Yorkshire, England, where a Bureau of Biotechnology provided consultant and analytical services to local fermentation industries. The early development of the principal biotechnology industries – food, drugs and textiles – display marked similarities, though their modes of development have differed over the centuries.

All began using empirically derived artisanal labour-intensive technologies which, following different patterns, were gradually mechanised, human hands being replaced by mechanical devices and machines driven, successively, by animal power, water, wind, steam engines, fossil fuels and electricity.

For many centuries, technologies and industries that processed foods and drugs have been closely associated. Standards of quality and safety for foods and drugs are frequently administered by the same regulatory agency, the US Food and Drug Administration being a typical example.

Beginnings of Biotechnologies

Sometime around 3,000 BCE the Chinese began acting on their belief in a close association of foods with medicines, both essential to good health, both at the time derived from plants and animal organs. The Chinese continue to believe that many ailments can be cured by appropriate diets; they were the first to add burnt sponge, an aquatic source of iodine, to people suffering from goitre. (The present fashionable addiction to 'nutriceuticals' is simply an extension of an ancient Chinese doctrine).

The earliest drugs were discovered by perceptive observation, trial and error, certain plants and their extracts found to alleviate pain and particular symptoms of disease. Some medicines still prescribed were discovered more than 4,000 years ago. Ephedrine extracted from a Chinese shrub *Ephedra sinica* was then and is now prescribed to relieve asthma and other allergenic respiratory dysfunctions. Between 4,000 and 5,000 years ago Egyptians prescribed remedies for more than 1,000 ailments. The Egyptian Ebers Papyrus [*ca* 1500 BCE] lists over 800 drugs compounded from such vegetable extracts as gentian, castor oil, vermifuges (anthelminthics to treat intestinal parasites) and henbane. The Papyrus prescribes treatments for rheumatism, schistosomiasis and diabetes.

A characteristic common to food, drug and textile industries is that empirically discovered and derived technologies long preceded any scientific understanding of the biochemical and biophysical properties of the raw materials and processed products. For many centuries food and textile industries progressed through mechanics, the gradual replacement of human hands by machines.

The principles underlying industrial processes of food fractionation and transformation, the spinning and weaving of natural fibres, were discovered several thousand years ago. Progressive mechanisation, which at first led to unemployment and deprivation among rural artisans, has significantly economised on human effort and energy expenditure in factories, restaurants and homes.

Food Preservation

The basic principles of food preservation that involve control of a. active water content, b. ambient atmosphere and temperature, c. pH, d. thermal inactivation of microbial and biochemical deterioration, were discovered long ago. Aboriginal, Asian and Mediterranean people used sun-drying to preserve milk, meat, fish, fruits and vegetables; sliced potatoes were freeze-dried by early Amerindians, from which the ice crystals sublimed in the dry air and low atmospheric pressures of the high Andes. Stone-age Britons dried moist grains over open fires to prevent sprouting. More than 4,000 years ago the Chinese preserved fish by osmotic dehydration with salt. Republican Romans reduced water activity in meat and fruits by adding salt or honey. About 5,000 years ago Middle Eastern farmers stored grains in filled earthenware amphora, hermetically sealed by impervious goat-skins, causing all stages of insect metamorphosis to be asphyxiated by respired carbon dioxide. Seneca describes how Romans preserved prawns and oysters in snow collected from the Apennines.

The modern frozen food industry developed after an American, Clarence Birdseye, observed how Aboriginal people in northern Canada preserved surplus whale, seal and reindeer meat in ice collected during the cold winter. Modern canning, bottling and boil-in-the bag were anticipated in Republican Rome by chopped spiced meats sealed and boiled in the cleaned womb from a sow or the body cavity of a squid. Preservation by fermentation and pickling was practised by Egyptians and other ancient Mediterranean people. Babylonians preserved all their milk by lactic fermentation. Fermented bread, beer and wine were produced on a factory scale during successive Pharaonic dynasties, while ethanol was distilled in China 3,000 years ago.

FROM HAND LABOUR TO MECHANISATION

Textile Technologies

The patterns and pace of mechanisation moved differently among the biotechnologies, textile industries progressing faster than food processing. Spinning and weaving of fibres from cotton, wool, flax and silk were practised 6,000 years ago in Egypt and Mesopotamia, the yarns being spun from a distaff, a cleft stick, onto a spindle. Spinning being exclusively an occupation of women, the word 'distaff' has long been associated with femininity. With relatively simple equipment, more than 4000 years ago Persians wove wool and cotton for clothing, luxurious silk fabrics, ornate woollen carpets and tapestries. Evolution from domestic spinning and weaving in rural homes to mechanised factory operations in Britain took many years. The compulsion to assemble workers and machines in factories arose from the need for cheap energy and to enforce more consistency in quality and uniformity. Until horse-drawn barges along canals carried cheap coal to energise steam engines, textile factories, in common with grain mills, were positioned on river banks driven by power transmitted from water wheels.

Textile technologies were mechanised by a series of inventions: John Kay's flying shuttle patented in 1703, various spinning machines invented in the late eighteenth century. From the Middle Ages rural women had spun wool and flax with the Saxon spinning wheel. Crompton's mule, invented in the 1760s, enabled one spinner to control 1,000 spindles, thus casting countless rural women and their families into unemployment.

None of the men who invented textile machines and none who contributed to invention and development of the nineteenth century steam engine were scientists or professional engineers. Thomas Newcomen was a plumber whose steam engine, designed to pump water out of coalmines depended on a piston driven by steam injected into a cylinder. By more accurate boring of cylinders that housed the piston and other devices, James Watt, an instrument maker, improved the efficiency of Newcomen's engine. One of Watt's early engines drove machines in a London flour mill, others gradually replaced water power in textile mills.

Ancient Egyptians dyed fabrics with plant pigments and the rich purple from shells of the aquatic murex snail discovered in the Mediterranean by the Pheonicians. During the Dark Ages the art of dyeing was lost among Europeans but was revived in the late twelfth century when Flemish dyers rediscovered plant pigments, their most spectacular being pigments from the roots and petals of madder (*Rubia tinctorum*) and the Asian lichen orchis (*Roccella tinctora*). From the mid-nineteenth into the twentieth century chemistry provided the greatest impetus to textile development, first through German synthetic aniline dyestuffs, later through synthetic fibres; viscose rayon regenerated from chemically modified cellulose was among the first.

Since the eighteenth century inventors and engineers have designed faster textile machines, with higher spindle speeds, direct winding onto looms and, most recently, fabric and garment design and manufacture controlled by computers. Over the past 50 years scientists have discovered the chemical composition of natural fibres, the changes that take place during processing, subsequent wear and tear, and from environmental stresses. Mathematicians have contributed to physicochemical analysis of the complex structures and changes involved in garments made up of millions of fibres largely held together by frictional forces.

Plant and animal scientists breed for improved fibre quality in cotton and sheep wool, chemists synthesise new fibres and fabrics, computer specialists and mathematicians devise programmes to design fibres, fabrics and garments and to control the machines that produce them.

Textile biotechnologies have developed from simple domestic and artisanal crafts into scientifically controlled and highly mechanised integrated industries.

But as with most industrial biotechnologies, progress to the present state took place over 7,000 years; innovations were devised by people who perceived opportunities for their inventive talents. Development proceeded spasmodically and episodically, knowledge sometimes passed on between generations, sometimes lost and then rediscovered. As disturbingly described by social historians and nineteenth century novelists, the pursuit of profit through mechanisation of textiles and other industries inflicted immense hardship and suffering on men and women displaced by machines. The horrors of the Industrial Revolution illustrate the persuasive need to assess systematically the social consequences of all development programmes.

Grain and Oilseed Milling

Milling of grain was the first food process to be mechanised, which is not surprising since grain fractionation by hand pounding and winnowing is the most tedious and time consuming of all food processes. Fractionation of grains by sieving and winnowing, extraction of olive and linseed oil by pressing was practised in Egypt and Mesopotamia more than 7,000 years ago. Commercial bakeries and breweries existed in Egypt and Mesopotamia 5,000 years before Eduard Buchner and Emil Fischer discovered the enzymatic conversion of carbohydrates into ethanol and carbon dioxide.

The earliest cereals to provide food for Mediterranean people were probably the seeds of wild emmer and einkorn, the outer hulls of which were tough, indigestible and so firmly attached that grains had to be parched over hot fires before being hand-pounded between a concave rock and a stone club, the earliest ancestor of the pestle and mortar. Some time around 8,000 BCE, naked wheat grains, free of outer hulls, the probable outcome of promiscuous transgenic hybridisation of emmer, einkorn and other grasses, were adopted and eventually cultivated [Hutchinson 1974; Simmonds 1976]. Primitive pestles and mortars were replaced by saddle stones, an early Egyptian device with which a kneeling slave propelled an upper stone with a lower smooth flat surface to crush grains spread over the concave surface of a lower stone slab. The shearing action to break open the grain was gradually improved by incising a herring-bone grooved pattern onto the working surfaces of the upper and lower stones. Later, attaching a wooden handle to move the upper stone backwards and forwards, and an eye-hole in the upper stone permitted a slave to feed grain continuously while moving the upper grinding stone. Grain milling thus became the first mechanised continuous food technological process, raw material fed and end product emerging in continuous flow.

Republican Romans adapted the Egyptian rotary quern in which an upper stone (the catillus) rotated over a lower stationary stone (the meta), grain being fed continuously via an eye-hole in the centre of the upper stone. Milled grain moved to the periphery by centrifugal force of rotation. Through a geared drive shaft, Roman querns were powered from water wheels. The first windmills were devised during the tenth century CE in what is now Iran and Iraq. Mills driven by water and wind power were dominant well into the nineteenth century; it is less than 100 years since grain mills powered by steam engines in North America were equal in number to mills driven by water wheels. In 1784 an early model of James Watts' steam engine was installed in a London wheat flour mill; the first mill driven by electric motors was constructed in 1887 in Laramie, Wyoming.

Though more precisely engineered than their ancient ancestors, modern grain mills equipped with incised break rolls are direct descendents of the Egyptian saddle stone and the Roman quern. Hand winnowing has its modern counterpart in an enclosed vibrating gravity table, combined with screens of various mesh size to fractionate particles of different dimensions. An ingenious software programme, devised in Bangalore, allows modern mills in India to be operated and controlled from a desk-top computer, while among poor rural Indian communities local grains are still ground by large wooden pestles and mortars, by saddle stones, and groundnut oil is extracted by animal-powered rotary querns.

The protein content of wheat, oilseed and grain legume flours can be enhanced by fine grinding and air-classification. In cereal, legume and oilseed grains starch granules, of varying size and effective mass, are embedded in a protein matrix. Fine grinding by dynamic shearing devices cause some starch granules to separate from the protein, which is shattered into fine fragments.

By air classification, in which a mechanical centrifugal force is opposed by a predetermined centripetal air drag, the lighter protein fragments and the heavier starch granules move in opposite directions. The starch granules in grain legumes and oilseeds being larger than in cereals, larger protein shifts can be produced from the former than the latter.

Science Applied to Biotechnologies

As with stone tools, biotechnologies evolved through three overlapping periods. The ancient empirical period began at least 10,000 years ago and evolved until the early eighteenth century. From then the middle period progressed as greater scientific understanding of biologies and biotechnologies gradually emerged. The modern Genomic Era began in the early 1970s. Systematic scientific study came late to industrial biotechnologies. The early history of food science, which progressed spasmodically during the eighteenth and nineteenth centuries, consisted of a gradual revelation of the composition and properties of food materials. Not until the twentieth century were novel food and textile processes devised directly from scientific principles. In contrast, during the nineteenth century pharmaceutical industries developed from the application of chemical knowledge learned in the German dyestuffs industries. Historical reviews of biotechnological progress are in earlier publications [Hulse 1995, 1996, 1998, 2003].

Evolution of Pharmaceuticals

Primitive people searched for natural panaceas and palliatives to cure their diseases and relieve their discomforts. Most early drugs were extracts of wild plants, some animal organs and a few available chemicals. Sumerian cuneiform tablets from the time of Hammurabi describe hepatic and febrile diseases, gonorrhoea, scabies, strokes and seizures. Babylonian drugs included extracts of hellbore, believed to cure psychological disorders, mandrake root and opium derived from the latex of certain poppies (*Papaver somniferum*).

From earliest times, sickness was regarded as punishment for behaviour offensive to the gods; curative medicine often being superstitiously associated with magic. During the fourth and fifth centuries BCE the Greek school of Hippocrates proposed over 300 remedies, most being plant extracts administered orally or via other body orifices. Aware of magical associations and that drugs in excessive doses could be fatal, the Greek word ¢áñìáêïí (pharmakon) can be translated as 'medicine', 'drug', 'poison' or 'magic potion'. The Hippocratic oath demands that in practising their diagnostic and curative processes physicians and surgeons must 'do no harm'. Three hundred years after Hippocrates, Dioscorides, a Greek physician formulated over 600 remedies from plant and animal organs, his medicines being prescribed over the next 1,600 years. During the second century CE, Galen of Pergamum added more remedies, known as 'galenicals', to Dioscorides' *Materia medica.*

It is impossible to discover what useful drugs the alchemists may have discovered in their pursuit of the *Elixir vitae,* the elusive substance that would ensure eternal life. The alchemists wrote their reports and results in cryptic codes and obscure symbols in order to confuse their competitors.

During the Middle Ages, Arab scholars achieved remarkable progress in science and the arts, in alchemy and medicine. In the eleventh Century CE a Mesopotamian physician, commonly known as Avicenna (his actual name was Abu Ali ibn Saud), translated the published works of Dioscorides, Galen, the School of Hippocrates and other Greek physicians into Arabic. Avicenna and his fellow Arab scholars also translated the writings of Aristotle and Plato, as well as other Greek philosophers and authors into Arabic. Subsequent translations into English and other European languages made possible our present access to Greek medicine and literature. Arabian *Materia Medica* of the period describes more than 1,000 drugs derived largely from medicinal plants.

From Empiricism to Science

From the mid-1800s analytical chemistry, microscopy and cytology progressed impressively. Anton von Leeuwenhoek, a Dutch draper, with no training in optics or any other scientific subject, learned to grind glass lenses and constructed a microscope with a 30x magnification by which he identified protozoa (which he named 'animalcules'), described blood corpuscles and the cellular structure of fish tissues. Chemotherapy, using chemical substances to inactivate disease-causing pathogens by bactericidal or bacteriostatic actions, was stimulated by identification of microbial pathogens and chemical means of controlling them. Wöhler's conversion of ammonium isothiocyanate into urea demonstrated that naturally occurring organic substances can be synthesised from non-biological sources. The discovery of hormones extracted from endocrine and ductless glands and later synthesised, added another dimension to therapeutic medicine, to the gradual development of modern pharmaceutical industries.

Despite some 20 per cent of modern pharmaceuticals being derived from natural and genetically modified organisms, a lively consumer and commercial interest in traditional sources of natural medicines persists and may be growing. Ethnobotanical expeditions in tropical forests of the Amazon have delivered more than 10,000 species for pharmacological examination. A vast unexplored resource of plants and microorganisms exists across the planet. Of more than 100,000 identified species, fewer than 200 microorganisms provide substances applicable in food and pharmaceutical industries. Higher orders of terrestrial plants represent more than 65 per cent of the Planet's biomass but fewer than 6 per cent of identified species are cultivated commercially. Of 80,000 vegetative species believed edible, barely 15 provide 90 per cent of human caloric intake.

The avaricious depredation of tropical forests by loggers and planters of cash crops, the despoliation of sites of rich natural biodiversity, deprive humanity of vast unexplored sources of medicinal plants and other potentially valuable organisms. Sadly no international agency exists with authority to prevent such predations.

Synthetic Drugs and Chemotherapy

Around the middle of the nineteenth century a chemist in London, William Henry Perkin, was trying to synthesise quinine, a drug used to treat malaria, traditionally extracted from the bark of *Cinchona* species indigenous to South American forests. Legend has it that, while heating a mixture of aniline and potassium dichromate, Perkin broke his thermometer, the liberated alcohol reacted with other substances to produce a rich purple pigment which he named 'mauveine', believed to be the first of the aniline dyes. Based on this result von Hoffman,

Perkin's tutor, started the German dyestuff's industry. Encouraged by success with synthetic dyes, German companies Bayer, Hoechst and Merck began chemical synthesis of drugs, first by producing analogues and derivatives of therapeutically active substances extracted from medicinal plants [Hulse 2004].

In the seventeenth century an inquisitive English physician discovered that rural people in the midland counties drank infusions of willow bark (*Salix spp*) to alleviate arthritic and rheumatic pain. The active principle in the bark was later analysed and identified as salicylic acid. Chemists at the Bayer Company reacted acetic anhydride with salicylic acid to produce acetylsalicylic acid, the first proprietary synthetic drug 'Aspirin'. Other synthetics, such as codeine produced by methylation of morphine, soon followed. In 1936 the British Medical Research Council defined 'chemotherapy' as medical treatment by synthetic chemical compounds that react with specific infective organisms.

Antibiotics

While engaged in microscopic studies to discover what microbes were devastating silkworms, Pasteur suggested that microorganisms might be induced to attack other microbes. In 1928 Alexander Fleming in London observed that a mould spore from *Penicillium notatum* accidentally infected and inhibited growth in a bacterial culture. The therapeutic potential of Fleming's observation was ignored until re-examined in 1939 by Florey and Chain at Oxford University. The effective bacterial inhibitor was isolated, chemically characterised and named 'Penicillin', the first of a long series of antibiotics subsequently extracted from cultures of various *Actinomycetes* species. Long before Fleming's observation, anthropologists reported how Micronesians scraped mould from tree bark which they rubbed into wounds to prevent festering [Hulse 2004].

Hormones

More than a century ago Claude Bernard, a French physiologist, reported how some critical bodily functions are regulated and controlled by 'centres of internal secretion'. Later the active agents were identified as endocrine and ductless glands that secrete hormones, the name derived from çiñìíí (hormon) a Greek word that means 'to urge on'. In the 1920s, adrenaline, extracted from suprarenal animal glands, was chemically analysed and industrially synthesised. In 1921, at the University of Toronto, insulin essential to diabetic patients was isolated from Langerhens Islets present in porcine pancreas. In the 1980s other Canadian scientists synthesised an insulin precursor from a genetically modified bacterium. Thyroxine, generated in thyroid glands, and various other hormones have since been chemically and/or microbially synthesised. Avian and bovine growth hormones, synthesised by genetically modified bacteria, promote growth in farm animals and cultured fish and stimulate lactation in bovines.

Food and Pharmaceutical Industries

Recognising their importance to human survival, healthy growth and activity, it is not surprising that food and drug industries constantly expand and diversify to satisfy changing demands of expanding, affluent and ageing populations. The total world value of industrially processed foods is estimated at more than $3 trillion, the sales value of commercial pharmaceuticals, excluding veterinary medicines, exceeds $500 billion, of which 49 per cent are sold in the USA, 24 per cent in the EU, 16 per cent in Japan. During the past 40 years food processing industries in India have grown to employ over 3 million people, to supply processed foods to more than 300 million regular consumers. The estimated value of foods processed in India in the year 2000 was more than 1,000 times the value in 1960.

Influence of Supermarkets

Supermarkets command exceptional influence over agricultural producers and food processors. Supermarkets, whose shelves may carry 10,000 distinct products and which supply some 90 per cent of all foods purchased by consumers in North America and the EU, exert pressures on growers and processors to meet the supermarket's quality standards, standards often difficult for small producers and processors to satisfy. Because of high sales volume, supermarkets can offer prices lower than small retailers. Though among poorer nations most food is sold from public markets, local stores and street vendors, supermarkets, many subsidiaries of European or American giants, are now rapidly expanding in Asia and Latin America.

Post-harvest Loss

To intellects of limited perspective, the primary purpose of food processing is to reduce what is variously described as 'post-harvest loss', 'degradation', 'deterioration', 'damage' or 'spoilage'. Post-harvest deterioration or spoilage, caused by biochemical change, microbial infections, insect infestations, attacks by other pests, chemical and other contaminations, can range from minor bruising or discoloration to foods heavily infected by pathogens or so spoiled as to be offensive and obnoxious. Post-harvest loss, of minor or major proportions, may involve biological, nutritional, economic or physical changes, result in lower weight, impaired appearance that is unacceptable, food unfit or unsafe to be eaten.

Assessments of post-harvest deterioration, degradation, damage, spoilage or loss, should state precisely what damage has been inflicted, the extent, cause, probable consequences and recommended means of remediation. Fully to understand the cause of post-harvest degradation it is essential to study systematically the entire sequence through which the product has been conveyed, to determine where, how and by what the damage was caused, a process that demands the competence of food systems analysts [Bourne 1977, FAO 1983, NAS 1978, Hulse 1995].

CHANGING PATTERNS AND FUTURE PROSPECTS

Research and Development

Though similarities between food and drug industries, and in lesser degree with textiles, are noteworthy, many divergent differences are apparent. Small-scale apothecaries that compounded medicines, dispensed pills and ointments have been driven out by huge drugstore conglomerates that dispense pre-packaged prescription and non-prescription drugs together with a vast assortment of toiletries and cosmetics.

Pharmaceutical industries that develop and manufacture drugs are composed of very few, large highly capitalised corporations, while food processors include giants such as Nestlé and Unilever as well as many thousands of relatively medium and small enterprises. Recent years have witnessed the creation of several thousand specialist 'biotechnology research companies', which, under contract or licence, develop new diagnostic, prophylactic and therapeutic substances and provide other technical services to pharmaceutical companies.

Pharmaceutical companies invest between 10 and 18 per cent of revenues in research, specialist biotech companies more than 30 per cent, and food processors an average of less than 0.5 per cent. A survey conducted in Canada a few years ago reported that of more than 3,000 registered food processors fewer than 20 maintained a liberally defined research activity. Only large food companies support a discrete research department, most others rely on raw material and equipment suppliers to provide technical advice. Some governments support national food research laboratories, which may offer technical advice to small industries. Many government research institutes cannot maintain a detailed awareness of specific opportunities, resources and constraints among hundreds of small and medium food companies. One encounters government food research institutes where few if any of the technical staff have experience in food processing industries. Consequently, while their scientific knowledge may be impressive, they possess little understanding of market forces, and opportunities. They therefore devise new products for which no consumer demand or profitable markets have been identified.

Biomedical Industries

The biomedical industries are developing into two inter-related entities: a. large pharmaceutical corporations, many of which operate internationally; b. specialist bioresearch enterprises, sometimes described as 'second generation biotechnology companies' or more simply 'Biotechs' or 'Biotech/Biopharm'.

Biotechs started about 30 years ago by using genetic modifications to synthesise proteins and derived drugs. In 1970 two Americans, Cohen and Boyer, demonstrated how using restriction enzymes DNA could be cut and transferred between unrelated organisms. In 1976 Boyer with several of his Californian colleagues created Genentech, now described as the first modern biotechnology company. Optimistic speculation that genetic transformations would quickly generate valuable new diagnostics and therapeutics encouraged enthusiastic investment in the new biotechnology enterprises. When Genentech made its first public stock offering its share price rose from $35 to $89 in twenty minutes, the fastest rise in American stock market records. Subsequently the fortunes of Genentech and other new biotechs fluctuated as investor expectations waxed and waned. Some failed, many survived: around the world there are over 4000 biotech/biopharm research companies. Over 70 per cent of Canada's so-called biotech/biopharm research companies were spun off from universities.

During their early years biotech/biopharm companies were relatively small and confined to laboratory syntheses and micro-processes. Some in North America have diversified to engage in preclinical and early clinical trials of potential new drugs. Among Canadian companies that describe themselves as Research Biotechs, 75% are small with fewer than 50 employees, 10 per cent are relatively large with over 150 employees. More than 50 per cent are engaged in the human health sector (therapeutics and diagnostics).

A growing group specialises in bio-informatics: data bases and information that they sell to other companies. Some provide analytical services, others molecular libraries from which samples of particular organic composition can be supplied to clients. Since they began as small bioscience research companies, Biotechs have diversified into specialised diagnostics, biopharmaceuticals, food and agriculture, renewable energy, environmental protection and clean-up, protection of biodiversity, bioinformatics and biotechnology systems.

Many Biotechs spun off from universities have failed. Some lacked industrial management and marketing experience. Others began with too little working capital to survive the long R & D period before a profitable product or service is developed.

Drug Development

The sequence of modern drug development begins by identifying targets for therapeutic intervention. These may include cellular or pathogen-specific molecules such as proteins critical in a targeted disease. Compounds may also be discovered from several thousand accessions in molecular screening libraries. Identified compounds are refined and subjected to a battery of biochemical analyses, simulation models and tests with laboratory animals. Compounds that pass preclinical screening proceed to a series of controlled clinical trials with human subjects. The high costs of clinical trials are forecast to increase at over 12 per cent a year. Drugs that pass clinical trials are presented for evaluation and approval by regulatory agencies. Highest risk and rate of failure occur during early preclinical evaluations. In North America only one out of several 100 compounds initially examined survives clinical trials; only 20 per cent of those rated clinically acceptable find profitable markets.

In December 2004, patents on some 15 highly profitable drugs expired. Consequently there is pressure on biotech/ biopharm and pharmaceutical companies to discover and bring to market effective new drugs of wide potential application. Pressures on drug companies are further exacerbated by rising demands from ageing North Americans and Europeans for pharmaceuticals, prosthetic devices and other healthcare services. Drug companies are on the horns of a dilemma: people suffering from serious maladies, such as HIV/AIDS, demand rapid approval of new therapeutics. Fearing costly litigation when new drugs cause unforeseen adverse effects, drug companies prefer to proceed cautiously and patiently until all clinical trials are satisfactorily complete.

Several Biotechs that began specialising in genomics, combinatorial chemistry, specific drug discovery or monoclonal antibodies to be licensed to larger companies, now carry new discoveries to pre-clinical, some to early clinical trials. Successful companies have defined their objectives according to their resources and concentrated on specific drugs for widespread diseases, or on technological systems ('platforms') amenable to several applications. Some specialise in diseases related to specific genes and means to remedy genetic dysfunctions.

To gain an adequate critical mass of resources, biotech research companies are acquiring or merging with competitors. Others enter into alliances with pharmaceutical corporations, some by mergers, others through formal licensing arrangements. Over one-third of American biotech research companies depend on licensing products and processes to larger biotech or pharmaceutical companies.

The Human Genome

Elucidation of the human genome sequence stimulated investment in pursuit of diagnostics and therapies for genetically induced maladies. Early expectations were modified when it was discovered that the human genome contains about 30,000 genes – not, as originally believed, close to 150,000 genes. Each active gene may encode for several different protein molecules, so there may be 5 times as many drug targets as there are genes. Recent speculations variously suggest that of the present estimated total of *ca* 500 targets, genomic research will lead to eventual identification of between 3000 and 10,000 targets, corresponding to proteins encoded by between 10 and 25 per cent of identified human genes. Following on Dolly the sheep, there is concern that knowledge derived from the human genome sequence could be adapted to clone humans, a practice some would equate with man trying to play God.

The ability to sequence genomes has stimulated an intriguing diversity of biotechnological sub-sectors. *Genomics* includes the study of the structure of genomes, genomic functions related to specific diseases. *Metabolomics* is devoted to the investigation of how metabolites influence physiological dysfunctions. *Pharmacogenomics* studies genomic and genetic patterns and malfunctions related to specific diseases, to molecular structures as they influence pharmacological activities. *Proteomics*, a rapidly expanding sub-sector, is concerned with specific proteins as they relate to particular infections, to pathogenic modes of action, proteins as carriers for vaccines, genes that encode for novel enzymes. Scientific institutions in some 18 nations are collaborating members of the Human Proteomics Organisation (HUPO), coordinated from a headquarters located at McGill University in Montreal.

In a recent lecture, Dr Elias Zerhouni, director of the US National Institute of Health, described how biomedical research and applications are changing from design and synthesis of pharmaceuticals to alleviate diseases to interventions at the molecular and cellular levels that aim better to understand the cause and etiology of various diseases, some communicable and others brought about by genetic dysfunctions.

Nano-sciences

Nano-sciences study the molecular constituents of biological substances and materials. 1.0 nanometre (nm) = 10^{-9} metre; the diameter of a human hair ranges between 100,000 and 200,000 nm. Novel analytical instruments, including the Scanning Tunnelling Microscope and the Transmission Electron Microscope facilitate determination of the structure of molecules. The properties of biochemical macromolecules are significantly influenced by the nature and structure of their constituent molecules. It is postulated that, by designing and synthesising novel molecules of desirable structure and properties, the functional behaviour of many proteins, pharmaceuticals and eventually food materials can be made more efficient and more effective [Roy Soc, 2004]

Costs and Revenues

Between 2001 and 2002 global revenues of Biotechs rose from $35.9 billion to $41.3 billion, of which 73 per cent was earned by American companies. In 2001, total revenues of the six largest bio-research companies amounted to about $8 billion, their research and development investments were between 20 and 30 per cent of revenues. These specialist Biotechs devise and develop new drugs and processes which some carry to preclinical stages of safety and effectiveness, and pilot-plant production scale. Larger pharmaceutical corporations expand the smaller biotech processes and subject the products to more intensive and extensive *in vitro* and *in vivo* clinical trials to determine potency, reliability and safety before submitting the new drug, with all available relevant scientific evidence, to regulatory authorities for official approval. Pharmaceutical companies are often criticised for their high prices and revenues. Critics ignore the exorbitant cost of developing a successful new drug, now in excess of $800m increasing at over 10% year, the time required from early laboratory synthesis being up to 15 years. Fewer than 1 per cent of substances initially synthesised are eventually approved, most being rejected at various stages along the way. The patent on a new drug lasts for only 20 years from the time the patent is approved. As soon as a patent expires, companies that are designated generic companies start to produce and sell the drug. Generic companies can sell at a lower price, since they incur no lengthy research and development costs. Some regulatory authorities have established departments specifically responsible for the monitoring of drugs processed by generic companies, to ensure that the generic version is equal in composition, potency and safety to the original.

To economise, several pharmaceutical companies have merged with competitors, others are 'downsizing'; Pfizer will soon close seven factories and terminate several hundred employees.

Bio-informatics

An early seventeenth century scholar complained: "The disease of this age is the multiplicity of books; we cannot digest the abundance of written matter." How to record, classify and provide access to the overwhelming volume of bio-scientific knowledge now generated deserves serious, systematic attention. More than 350 bio-informatic databases operate around the world. Urgently needed are bio-scientists and bio-philologists competent to compile and maintain reliably accurate databases and bio-information systems using precise terminologies comprehensible to all concerned. Database curators need competence to recognise accuracy and relevance, to delete outdated accessions, and interpret needs of potential users. Several companies maintain molecular libraries that offer computerised molecular structures and provide samples of the substances illustrated in their computers to research biotech and pharmaceutical companies. An urgent need of bio-informatics is for an internationally accepted thesaurus and dictionary of precise definitions. The penchant to coin new names without stating precise definitions or how they may conflict with existing terminologies persists as a serious impediment to rational databases and cogent bio-scientific communication.

TRENDS AND PROSPECTS

Semantic Confusion

Any attempt to suggest how the future of industrial biotechnologies may unfold is hampered by confused concepts and imprecise interpretations of what is connoted by 'biotechnologies', with no clear distinction between 'ancient' and 'modern' technologies. Disturbingly typical is a publication by the Directorate-General for Press and Communications of the European Commission [Eurobarometer 2003]. The publication is the fifth in a series of reports of surveys to assess the attitudes of Europeans to 'biotechnology'. The report speaks of 'modern biotechnology' without any definition of its intended meaning. It differentiates between 'GM Crops' and 'GM Foods' seemingly unaware that many GM foods are derived from GM crops, some with ingredients produced by GM organisms. GM foods are described as 'using modern biotechnology in the production of foods, for example to make them higher in protein, keep longer or change the taste'. GM crops are described as 'taking genes from plant species and transferring them into crop plants to increase their resistance to insect pests'. Foods may be preserved, processed or transformed by modern biotechnologies that do not involve genetic modifications. Is a food partly processed by an enzyme, or with a flavour or micronutrient derived from a GM organism a GM food? When does a modern biotechnology cease to be modern?

Biotechnologies employ a diversity of cell cultures. Various metabolites applicable in food processes are synthesised and extracted from cultured cells of microorganisms, higher plants and other organisms that have not been transgenically modified. The EC report states that one-half of the people surveyed were asked for their opinions on 'biotechnology', while the other half were questioned about 'genetic engineering'. In the final analysis the results from the two groups were combined, clearly indicating that in the minds of the surveyors 'biotechnology' and 'genetic engineering' are one and the same. By rational scientific interpretation, 'genetic modifications' are simply components of a very broad and constantly expanding spectrum of 'biotechnologies'.

Over the past 20 years, biotechnologies have evolved from intellectually intriguing biosciences, into industries that process and produce biological products from bio-catalytic reactions, from natural and GM microorganisms, plant cells, mammalian and insect cells. Some techniques modify genetic composition and expression, others accelerate and adjust metabolic processes.

Cell cultures complement chemical modifications in syntheses of diagnostics and therapeutics. Cultured microbial and viral cells, some genetically modified, produce a diversity of hormones, nutrients and drugs. Insect cells infected with adeno-and baculo-viruses produce proteins used in gene therapy and viral vaccines. Mammalian cells synthesise therapeutics of superior purity: monoclonal antibodies, viral vaccines and Factor VIII to promote blood clotting in haemophiliacs.

Cell cultures demand qualified bioengineers and systems analysts to rationalise interactions among many complex variables in the selection of media, variable conditions of culture, all of increasing complexity as culture processes progress from laboratory, through pilot plants to factory-scale.

Pathogenic bacterial colonies transmit biochemical signals to determine when they have sufficient umbers to overcome a target host's immune defences. They then turn on virulent genes to become potently infective. Scientists in Wisconsin have developed derivatives of L-homoserine lactones to block bacterial signals rendering the pathogens susceptible to antibiotics.

Molecular Modelling and Virtual Reality

Pharmaceutical industries began by extracting, screening and chemically modifying pharmacologically active substances of natural origin. These traditional processes are giving way to identification of how specific diseases are caused, the mechanisms by which particular drugs act to diagnose, prevent or cure them. The latest generation of prophylactics, therapeutics and diagnostics are designed by molecular modelling in programmed computers. Late in the last century, using classical organic chemistry, a scientist could synthesise about 50 new active compounds in a year. Today, computer assisted biosyntheses can generate several thousand new compounds. Programmed computers devise molecules to be systematically compared with molecular structures stored in the computer memory or downloaded from molecular libraries. In less than a year a million compounds can be screened against specific target proteins.

Other biotechnological innovations include 'Genomics' – the study of genomes and DNA nucleotide sequences; 'Proteomics' that study specific proteins produced by genomes; 'Metabolomics' that study the influence of gene expression on metabolites synthesised in cell cultures.

'Virtual Reality' provides novel means to study molecular structures, DNA, whole organs, plant and animal cells. At the University of Calgary, Java 3-dimensional technology projects images onto the walls and floor of a cubic room 2.5 × 2.5 × 2.5m. The observer virtually stands in the middle of the molecule or human organ being studied. Images examined through specially designed spectacles can be expanded or moved around by a manual control. Trained observers can study 3-dimensional structures of DNA to discover aberrations or unusual features; the structure of whole organs; the structural relation between a synthesised diagnostic or therapeutic and a diseased cell.

Downstream Processes

In biotechnological nomenclature, 'downstream' relates to all processes that follow synthesis in a bioreactor: fractionation, isolation, purification and sterilisation of synthesised products. Most difficult and expensive are scale-up processes: expansion from a laboratory flask, through pilot plant to processes of industrial scale. In North America, downstream processes absorb some 80 per cent of factory production costs. Highly skilled bioengineers are in urgent demand to improve the efficiencies and economies of downstream processes. Recent studies of skilled human resource needs indicate that present and predicted demand for biotechnologists and bioengineers significantly exceeds expected supply.

Separation and Isolation

Different primary products are isolated from different fractions: insulin from harvested cells; several vaccines from supernatants; intracellular metabolites are released by rupture of cell walls. Substances sensitive to excessive heat, organic solvents and other chemicals, need benign systems of fractionation, isolation and sterilisation. In bio-chemicals sensitive to heat, microorganisms can be inactivated by high-temperature-short-time [HTST] sterilisation, or removed by microfiltration. Intact cells can be harvested by continuous centrifuging, but carefully controlled conditions are needed for sensitive mammalian cells. Cell disruption to isolate intracellular metabolites can be mechanical, chemical or enzymatic. Mechanical methods include high-pressure homogenisation and maceration using small glass or plastic beads. Enzymatic lysis of cell walls are conservative of energy and less prone than mechanical or chemical methods to degrade sensitive bio-synthates. Processes catalysed by enzymes, many derived from GM organisms, being less demanding of energy and more benign than many chemical catalysts, are finding increasing bio-industrial applications.

Optimum conditions for isolation and purification of metabolites vary with the substances involved. Liquid extraction is useful for antibiotics; fractional distillation for heat-tolerant organic chemicals; freeze-concentration for heat-sensitive substances in aqueous dispersion, aqueous-phase liquid fractionation for heat-sensitive enzymes. Membrane separations include reverse osmosis, ultrafiltration, microfiltration, electrodialysis and nano-filtration. Ancient Egyptians applied microfiltration to purify water from the Nile that slowly percolated from settling ponds through fine sand filters.

At pressures between 10,000 and 40,000 kPa carbon dioxide behaves as an organic solvent and is used to extract bio-chemicals intolerant of organic solvents. Supercritical gas and liquid extraction [SGE] is used industrially to isolate essential oils and oleoresins from aromatic plants, and caffeine from coffee.

Processes of Preservation

Most foods and pharmaceuticals are susceptible to deteriorative change: through microbial infection, pest infestation, biochemical reaction, chemical contamination and atmospheric pollution. Sensitive foods and pharmaceuticals can be preserved by processes that

a. inhibit, destroy, remove or prevent entry of harmful microorganisms

b. restrict adverse biochemical and biophysical change.

Biochemical spoilage induced by microbial infection or catalysed by enzymes can be restricted by

a. thermal destruction or inactivation of an offensive microbe or enzyme

b. reducing temperature of the product and ambient atmosphere to below freezing

c. lowering the active water content (i.e., product vapour pressure) by partial dehydration or elevation of soluble solids

d. control of gaseous ambient atmosphere within the package or container.

Reducing the Active Water

Active water [A_w] is defined as the quotient of the vapour pressure of a particular biological material divided by the vapour pressure of pure water. Reducing Active Water restricts microbial growth and biochemical degradation since most enzyme catalysed degradations require an aqueous environment. Active water may be reduced a. addition of soluble solids, the A_w of saturated salt or sugar solutions are significantly lower than the A_w of pure water; b. removal of constituent water by processes of dehydration.

Partial dehydration of harvested cereals, meat, milk, fruits and vegetables by radiant heat from open fires or by exposure to direct sunlight has been practised for many hundred of years. More benign dehydration systems, that inflict less thermal damage, rely upon shorter times of exposure to heat, evaporative cooling, lower temperatures and reduced ambient pressures to accelerate rates of aqueous evaporation. Dehydration under vacuum depresses the boiling point of water and reduces ambient resistance to evaporation.

Lyophilisation [freeze drying] is protective of biological materials highly sensitive to thermal damage. Most biological cells are ruptured by large ice crystals that form during slow freezing. Lyophilisation begins with rapid cryogenic freezing by immersion in a medium below-150°C, ice crystals, rapidly formed are small and do not rupture organic cells. The frozen material is enclosed in a vessel sufficiently robust to withstand an internal pressure below 5.2 mbar under which conditions the material remains frozen. The constituent ice sublimes directly into water vapour which may be captured on a condenser. Rates of sublimation can be accelerated by black surfaced heating plates or by micro-wave heating.

Controlled Atmosphere Storage

Perishable foods, sensitive to oxygen or other ambient gases, may be stored and/or packaged under partial vacuum or an inert gas. Deterioration by oxidation can be reduced by storage/ packaging under partial vacuum, nitrogen or carbon dioxide. Many fruits synthesise and exude gaseous ethylene which accelerates ripening and reduces shelf-life. Bananas stored under an atmosphere containing 5 ppm of ethylene will ripen in 2–3 days. Storage under an atmosphere high in carbon dioxide and relatively free of ethylene will delay ripening for 16–18 days.

Thermal and Non-thermal Preservation

Thermal processing can impair critical functional and physiological properties in addition to inactivating undesirable micro/organisms and enzymes; the higher the temperature, the longer the exposure, the greater the degree of damage. Food processors are faced with a challenging dilemma. During the past 30 years, reported food-borne infections have increased by several orders of magnitude, pathogens such as *Campylobacter, staphylococci*, novel strains of *E. coli* and *Listeria* recently added to the long-time nuisance pathogen *salmonella*. Consumers now demand foods perceived to be healthier, fresher, and less perceptibly altered from their natural state.

Industrial technologies are designed to inactivate or slow down microbial and biochemical processes of spoilage and degradation. To restrict thermal damage, various relatively benign thermal and non-thermal processes are under investigation. Processes with short exposure times include spray-drying, tubular and scraped-surface heat exchangers, steam injection before aseptic packing. Aseptic systems rely on physical exclusion of microorganisms and contaminants from sealed sterile systems by which processed products can be conveyed or stored. Thermal damage can be restricted by uniform generation of heat for a relatively short time.

Ohmic heating is effective for biological materials sufficiently fluid to be pumped through a column across which an electric current passes between two electrodes. Heat is generated rapidly and uniformly, the sterilised product is rapidly cooled, aseptically conveyed and packaged in sterile containers. Commercial models range from 10 kW pilot units that process 100 kg per hour up to 300 kw machines with capacities of 3 tonnes per hour.

Microwave (MW) and *radio frequency* (RF) heating depend on electromagnetic energy from a magnetron that produces an electric field alternating at microwave and radio frequencies. Heat is generated by rapid reversal of molecular polarisation. MW and RF provide uniform short-time heating at high internal temperatures. MW and RF are used industrially in dehydration, microbial inactivation and rapid cooking, with optimum processing parameters determined by computer simulation models.

Non-thermal Processes

These include *high isostatic pressures* between 50 kPa and 1 MPa applied to biological materials high in water content. Products are sealed in evacuated flexible packages immersed in fluids; high pressures are uniformly transmitted through the fluid medium to the packaged materials. Ultra-high pressures cause temperatures to rise in most compressible substances, the degree of increase depending on the specific heat and other physical properties. Heat generated in compressed gases provides the operating principle of diesel engines. Materials high in water suffer measurable but lower temperature increases than compressed gases. Under adiabatic conditions near to normal temperatures, temperatures in high moisture foods rise in linear relation to the pressure applied, temperature rise among different foods ranging from 3° to 6° for every 100 MPa pressure increase.

Three forms of *pulsed energy* for microbial inactivation are under study: a. Pulsed electric fields (PEF); b. Pulsed light (PL); c. Pulsed magnetic fields (PMF). Electric potentials induced by PEF cause lethal polarisation of microbial cell membranes, empirically determined critical potentials varying among microbial species, cellular morphologies and ambient conditions. Alternating pulses seem more effective than constant polarities at field strengths between 15 and 30 kv/cm. Both PEF and PL, the latter generating light flashes at 20,000 times sunlight intensity at the earth's surface, are effective against vegetative microbial cells, but not against spores or active enzymes. Oscillating magnetic fields are said to inactivate vegetative microorganisms though the literature on industrially viable applications is sparse.

Ultrasonics at frequencies above 20 kHz, causing alternating pressures and cavitation, can inactivate vegetative microorganisms in liquid suspension. Ultrasonics may also be used as an adjunct to crystallisation, filtration, hydrogenation and ageing of alcoholic beverages.

Ionising radiations can inactivate microorganisms and destroy insect infestations. Radiation sources include gamma rays from such radioisotopes as ^{60} Co and ^{137}Ces; X-rays or electrons generated by machines. Absorbed radiation is measured in Grays or Kilograys (kGy), one Gray equivalent to one Joule/kg. In the United States irradiation is permitted to preserve various biological materials. Permitted maximum doses are indicated parenthetically for the following: fresh poultry (3 kGy), dehydrated enzymes (10 kGy), spices (30 kGy), pharmaceuticals (above 30 kGy dependent on biochemical nature and conditions of use). Higher levels are allowed in food ingredients consumed in small quantities (e.g., spices) and in prescribed pharmaceuticals.

Process and Product Quality Control

Simply defined, quality control (QC) objectives are to ensure properties of raw materials and final products comply with defined specifications and specified properties are consistently maintained in all products processed over time.

QC specifications are defined by: a. international protocols, b. government regulatory agencies, c. customers, secondary processors and retailers, d. food and drug processing companies.

For many years product QC was determined by extracting representative samples of processed products to be assessed and analysed in quality control laboratories. *Ex post facto* laboratory analyses are gradually being superseded by on-line sensors and monitors that continually assess and detect defects in critical properties. Feed-back signals correct faulty processing parameters, all on-line determinations being recorded in computers. Sensors variously depend on biological, chemical, biochemical, biophysical, microbial, electronic and photoelectric signals. They may record apparent viscosities, temperatures, pressures and relative humidity. Chemical sensors respond to pH, specific ions, organic radicals and impurities. Biosensors may employ immobilised enzymes, bacteria, antigen-antibody reactions and DNA probes. Spectroscopic methods, too many and varied to be comprehensively described, respond to physical and rheological properties and to many chemical constituents.

Magnetic resonance imaging (MRI), generally too expensive for food processing but used by pharmaceutical industries, depends on the different magnetic properties of atomic nuclei placed in a magnetic field that induces different energy levels between protons aligned with and against the field. MRI can detect many defects and being non-destructive, installed online it can assess 100 per cent of processed, packaged products, and is particularly useful in monitoring packaged pharmaceuticals.

On-line sensors must be accurate and reliably responsive, tolerant to processing conditions, easy to install and maintain.

Sensors and their responses must be systematically checked and correlated with direct analytical determinations. It is reported that up to 1000 different nano-sensors can be carried on a single on-line sensor-detector.

Final Comment

The foregoing illustrates how industrial biotechnologies have progressed from perceptive artisanal empiricism to scientifically devised and controlled systems. Scientific development and control are more advanced among pharmaceutical than among food and textiles industries, where scientifically devised systems continue side by side with traditional, empirical technologies. Supermarkets and other dominant retailers compel food and textile industries to deliver goods at ever lower prices. Low prices are attractive to consumers but present trends forebode that only large food industries can afford the research needed to develop and deliver products of consistent quality at prices demanded by supermarkets. Eventually this could result in significant reduction in the diversity of foods available and affordable to most consumers and the elimination of many small-scale farmers and retailers. Large department stores and supermarkets that seek to undercut the prices of competitors for textile goods are stimulating production by low paid workers, some in poor countries, others in oppressive sweat shops staffed by immigrants.

Given the high cost of modern biotechnology research and development, it is doubtful if the innovation, design, production and distribution of all necessary foods and essential drugs can be controlled solely by market forces. Government interventions are inevitable if all humanity is to be assured of food and health security. Mortality and morbidity from tropical diseases will be exacerbated by global warming; pandemics will spread across the planet by intercontinental migration of humans, birds and by goods transported in international trade. The need for progressive research to expand and refine biotechnologies is a *sine qua non*, for as René Descartes observed in his 17th century *Discourse on method*: "All at present known is nothing in comparison with what remains to be discovered."

ENVIRONMENT AND RESOURCES

The Planet's Creation and Evolution

The noun 'Environment' is derived from an ancient French word meaning 'to surround' or 'to encircle'. 'Environment' is defined in the *Oxford English Dictionary* as 'the surrounding conditions or influences under which organisms, persons or things exist and develop'. From this broad definition 'environment' is applied to various physical, political, economic, academic, social systems and conditions. This text is confined to the Planet's physical and biological environment, its atmosphere, climates, ecosystems and natural resources and their relation to biotechnologies.

Anthropological, cosmological and geological evidence indicates that from the time of its creation the Earth, its environment, ecologies, natural resources and the creatures and organisms that have been its inhabitants have continually changed, sometimes by gradual evolution, sometimes from extreme and disruptive occurrences: collisions with large asteroids and meteorites, violent earthquakes, volcanic eruptions and extreme climate change.

Justifiable is a growing concern with undesirable changes accelerated by human activities: gaseous and particulate emissions from combustion of fossil fuels and timber; agricultural, industrial and domestic pollution of atmospheres, inland and coastal waters; deforestation; degradation of land and ecosystems; deliberate or consequential depletion of biodiversity. A comprehensive review of how the universe and planet earth were created and have evolved, the origins and evolutions of living organisms is recorded in a special issue of *Scientific American* [*Scientific American 1994*].

An increasing number of reputable observers contend that the styles of more affluent middle classes, now expanding as countries such as the People's Republic of China and India rise in prosperity, with dilating demand for motor vehicles and other amenities, cannot be sustained with the resources available. Resources are being used at a faster pace than they can be restored. Human-induced modifications to the environment, to the global biochemical and hydrological cycles are rapidly accelerating. "Most transformations over the past 10,000 years have occurred during our lifetimes" [Kates 1994].

Governance of Environments and Resources

The United Nations Environment Programme, established following the 1972 Stockholm conference, conducts research and surveys related to environmental issues. Other UN agencies and international organisations study and report on the Planet's diverse resources, provide estimates of consumption and depletion, and recommend protocols, procedures and practices to encourage conservative resource utilisation.

Sadly no single international body possesses the mandated authority and responsibility to protect the Planet's environment, nor to control access to and consumption of its natural resources. Given the isolationist political ideology of the most powerful nation, creation of an international environmental authority empowered to enforce conservation seems improbable in the near future.

National, state and local governments can enact and enforce laws related to resources within their territories and jurisdictions, for control of atmospheric and other forms of pollution, for disposal of sewage, hazardous and other wastes. Governmental legislative protocols and regulations vary among nations, many persuasively influenced by industries and organisations with vested interests.

Conformity to such international conventions as relate to reducing atmospheric pollution, protecting biodiversity and overexploitation of marine resources is voluntary and no nation can be forced to ratify or comply with what is proposed. Among abstainers to convention conformity are the most egregious polluters and consumers, both governmental and industrial, of non-renewable resources.

Significant inconsistencies among nations are apparent anent ownership of and access to inland waters, natural forests and public land. Nor is there uniformity in how values of natural resource assets are assessed and accounted for. In some jurisdictions, trees have no declared value until they are chopped down; inland waters are free to everyone until they are pumped and piped to homes, industries and farmers. The failure to charge a rational price for water delivered to farmers leads to excessive consumption.

Timber companies that clear-cut natural forests to extract only a few prime species are rarely charged for the unused timber left to rot.

In several Latin American countries it was discovered that the trees extracted and used by lumber companies represented less than 15 per cent of the timber hacked down. There is no international authority empowered to force industries or individuals, whose activities inflict damage, despoliation or pollution on other nations, to reimburse their neighbours for the damage they suffer.

It would seem, therefore, that environmental protection and resource conservation are a collective, uncoordinated responsibility among individuals, communities, industries, state and national governments.

Ideally all parties would cooperate with counterparts in other nations, with sensitive concern for the Planet's resources and the welfare of all its inhabitants. Sadly those nations and communities that consume most, who cause greatest environmental disruption and resource depletion, seem least disposed to exercise constraint and prudent conservation; they assume an inalienable right to increase their excessive consumptions. Brundtland's plea to act with consideration for the needs of future generations falls more on deaf than receptive ears.

Are Resources Sustainable?

During the twentieth century world population grew from 1.6 billion to 6.1 billion. How rapidly and to what size it will expand during the twenty-first century defies prediction. In absence of some cataclysmic event or widespread disease epidemics, population growth, with persistent rising consumption of natural and industrially derived resources, will irreparably stress the Planet's resources and ecologies. Contradictions over potential consequences of ever-rising populations, resource depletion and environmental change are polarised between two extreme positions:

(a) increasing population and resource demands pose a serious threat to environmental stability and finite resources;

(b) human creativity and scientific ingenuity will devise technological means and resources by which to increase the Planet's resources and human carrying capacity.

One school contends that scientific study should focus less on population growth and concentrate on assessing present and predictable increases in *per capita* consumption of renewable and non-renewable resources [Arizpe *et al* 1992]. The British New Economics Foundation, based on sound scientific evidence, states: "The Planet's ecological stocks are being depleted faster than nature can regenerate them." It seems evident that the Planet's essential resources cannot sustain predictable future demand and that either the more affluent must reduce their patterns of consumption or the poorest will suffer intolerable deprivation. In light of the serious and growing incidence of obesity and the many resultant diseases, it would seem desirable that all physically able human beings would be wise to use bicycles or walk more frequently than to travel by automobile.

Confusion of Carrying Capacity

Various soothsayers offer disparate predictions of the Planet's human carrying capacity, that range from 7.5 billion [Gilliand 1983] to 40 billion [Revelle 1976] to 50 billion [Brown 1954].

The concept of carrying capacity derives from animal husbandry: how many grazing bovines or ovines can be sustained on any given area of natural or cultivated pasture. Such estimates of global carrying capacity are of little practical utility. People are not grazing animals; they live, move and have their being in nations and communities, among and within which access to resources vary significantly. A more equitable sharing of worldly wealth, power and resources among all peoples, while highly desirable, seems unattainable given present egocentric dispositions. Global carrying capacity would differ greatly if all communities adopted America's level of resource consumption or everyone on earth adopted more conservational standards. Observers described by some as gloomy pessimists by others as pragmatic realists, profess beliefs that non-renewable resources, ecosystems, biodiversity, goods and services essential to healthy lives are being depleted, degraded, despoiled and over-taxed beyond economic or scientific restoration.

Ecosystems are the basic functional units of ecologies. They comprise a mix of abiotic materials and organisms that produce, consume and/or decompose other organisms. When undisrupted by external activities and influences, components of each ecosystem evolve and develop to co-exist in a relatively balanced symbiosis. If one or two components in an ecosystem are destroyed or depleted the balance is upset, other interdependent organisms disappear, exotics may arrive, biodiversity is deliberately and consequentially altered.

At the end of the twentieth century affluent OECD nations, that represented 16 per cent of the world's population, consumed close to 75 per cent of gross global product and almost two-thirds of the Planet's energy sources. Among Asian nations almost all available arable land is under cultivation, quantities of cereal grains harvested *per capita* have virtually levelled off, urban populations grow at 3.5 per cent a year, rural people migrate into cities, motor vehicles increase proportionately faster than human populations. 'Carrying capacity' is therefore more complex than a simple arithmetic relation between population numbers and the size of potentially arable land.

The People that on Earth do Dwell

It is difficult to determine accurately the size of populations resident in any region or nation. Organisation of a national population census is costly and entails employment of many enumerators and statisticians. Complexities are compounded by shifting populations, migrations provoked by civil conflict, people seeking a better quality of life, by irregular dispersions of families and communities, some to places difficult of access. All quoted global demographic statistics must therefore be regarded as best estimates, biggest errors being among nations with least resources.

Blaxter presents an estimated pattern of world population development since settled farming began around the end of the Pleistocene period [Blaxter 1986]. Between 10,000 BCE and 01 CE the world's population is believed to have grown from 10 million to some 250 million; between 01 and 1800 CE from about 250 to 550 million. Population growth was never consistent or uniform; numbers were intermittently reduced by plagues and pandemics in the fifth and fourteenth centuries, by conflicts and famines in most centuries.

In 1950, world population was assessed as 2.6 times the size in 1800; from 1950 to 2000 estimated growth was from 2.5 billion to 6.1 billion. How rapidly and to what eventual dimension world populations will expand during the twenty-first century and whether global population will eventually stabilise is highly speculative. Over the near future, urban communities will expand and become more dense and congested, patterns of food and resource consumption will change, stress imposed on all natural and industrially processed resources will be more destructive and depletive.

Existing inequitable conditions, with the rich and powerful acquiring greater wealth and power, while the poor and powerless suffer more abject misery, will exacerbate violence. Communities forcibly driven from their homes, dispossessed of their properties, deprived of their livelihood and condemned to near slavery, are inevitably motivated to attack those they regard as the responsible enemies. Whether those who attack their predators should be castigated as terrorists or hailed as patriots and martyrs, fighting and dying to recover their rights, is of conflicting opinion. As illustrated by the excessive damage and deaths inflicted upon Lebanon during the summer of 2006, opinions are entirely influenced by governments and peoples who support one or the other side.

Population growth and life expectancy

Year BCE	World popn [M]	Year CE	World popn [M]
10,000	10	01	250
5,000	40	500	200
1,000	80	1000	250
500	150	1500	450

		Popn [Bn]			Growth	Annual %	Increase
	1950	**1990**	**1995**	**2003**	**1950–2003**	**1980–85**	**1990–2003**
World	2.52	5.28	5.92	6.24	×2.5	1.8	1.6
Africa	0.22	0.63	0.73	0.86	×2.7	2.9	2.7
Asia	1.40	3.20	3.46	3.73	×2.7	1.9	1.5
Latin America	0.17	0.44	0.48	0.55	×3.2	2.1	1.6
Europe	0.54	0.72	0.73	0.75	×1.4	0.4	0.1
Developing	1.79	4.27	4.67	5.14	×2.7		
Developing							
% of World	71	81	82	82			

	Life expectancy (years)		**<15yrs**		**Age groups % >65 Yrs**	
	1975	**2000**	**1975**	**2000**	**1975**	**2000**
World	58	66	37	29	6	7
Africa	46	49	45	44	3	3
Asia	56	68	40	29	4	6
Latin America	61	69	45	44	4	5
Europe	71	74	24	17	11	15

Europe, in common with other affluent regions, is home to increasing numbers of older men and women, a rising proportion of non-producers with greater demands for health care and social services. In contrast, nations of Africa, Asia and Latin America include higher proportions of active young men and women who need employment and have larger appetites. If their native countries are unable to feed and sustain them, there will be pressures on more affluent nations to permit larger immigrations from poorer nations. The populations of Africa's poorest nations are predicted to double during the next 50 years.

URBAN DEVELOPMENT

Christians and Jews, who accept the biblical accounts of creation as literal truth, believe the first humans appeared in the Garden of Eden. In contrast, Babylonians believed their god Marduk first created a city, which they called Eridu, and the first human beings to inhabit and rule it. Remains of Eridu were discovered by archaeologists in the mid-nineteenth century at a site known as Abu Sharein. The Judean-Christian mythology may have inspired the poet William Cowper to write:

"God made the country, but man made the town." Since no one has discovered the Garden of Eden, while Eridu was inhabited for several hundred years, Babylonians may have reason to claim theirs as a more credible version of human creation. Eridu was located on the south bank of the Euphrates at the edge of the fertile alluvial plain, close to the southern marshes.

The most enduring legacy of the Babylonians was their invention of printing in Uruk and the planning of cities, of which there were several dozen during the last 3,000 years BCE. Each city controlled its system of governance, surrounding arable and pastoral territories, water management and irrigation. By efficient crop cultivation on the fertile alluvial soil, with supplementary irrigation from elaborate systems of canals and dikes, fed from the main rivers, Mesopotamian farmers produced food in excess of their subsistence needs. Surpluses provided food for the growing urban populations, enabling them to engage in non-agricultural profitable commercial and administrative activities.

The invention of cuneiform script in the city of Uruk about 3500 BCE, which eventually evolved into a hybrid of pictographic and phonetic signs, provided detailed records of how the Mesopotamian cradle of civilisation developed. Cuneiform is a method of writing in which a stylus cut from reed is pressed into soft clay to form wedge-shaped imprints. An assortment of patterns and designs depict different words and numerals [Latin '*cuneius*' = wedge]. [Leick 2002].

The ten most important Mesopotamian cities began with Eridu about 5000 BCE, and continued until Babylon, which lasted from around 2000 BCE until its decline during the Sassanian occupation between 225 and 600 CE. (The name 'Babylon' derives from the Sumerian 'Bab-Ilanl' meaning 'Gate of the Gods'.) When Alexander the Great defeated Darius in 331 BCE he intended Babylon to be the capital of his Persian empire. Had Alexander not died prematurely in 323 BCE, Babylon not Baghdad might now be the capital of Iraq. Baghdad, the city created by Al-Mansour, ruling Caliph from 754 to 775 CE, was the capital of the vast Abbasid empire. The city is located between the Euphrates and Tigris where the two rivers flow closest to one another. It was officially named 'Madinat as-Salam', 'City of Peace' but is known as Baghdad, name of the small village it displaced and superseded (Lewis 1993).

Various influences have led to the growth of cities and urban development around the world. Industrialisation in Britain caused textile spinning and weaving in rural homes to be replaced by textile factories powered by cheap coal. Countless unemployed rural people migrated into urban slums. Rapid mechanisation of American agriculture forced similar rural to urban migrations in the mid-nineteenth century. In many lands, migrations of young people into cities are stimulated by the attraction of amenities for entertainment and diversion unavailable in rural areas.

Unplanned Urban Growth

Unfortunately, most contemporary cities, unlike those of ancient Mesopotamia, have grown rapidly with little planning or foresight. Older cities in North America and other territories now suffer from disintegration, breakdown, inadequate restoration and maintenance of infrastructures and services essential to healthy and agreeable living conditions. Atmospheric pollution from vehicle exhausts, factory chimneys and other fuel combustions causes erosion of ancient buildings and serious respiratory ailments, now the prime cause of morbidity and mortality in several large Indian cities. Disposal of urban waste, garbage, domestic and industrial sewage, contamination of air and water by pathogenic and obnoxious pollutants urgently need systematic monitoring to prevent insult and injury to human health; all need greater planned investments than are presently provided.

A major challenge, often ignored, is to deliver adequate, acceptable and safe food to urban markets and consumers. As cities grow and become more congested, the time to transport food from rural producers to urban processors and consumers inexorably increases. The longer the time taken to transport perishable crops, livestock and fisheries products, greater is the probability of deterioration, spoilage and waste. Pearson, Brandt and Brundtland all described accelerating rates of urban growth and the urgent need for foresight and planning. In several publications this author has advocated systematic analysis of urban-to-rural food distribution, analyses that make use of simulation models to compare alternative routes, times of day or night and modes of transport that provide rural-to-urban food conveyance in an efficient and economic manner. Rural-to-urban migrations could be slowed by providing rural employment opportunities in facilities for primary processing of food and other agribusiness activities [Hulse 1995, 1999].

Table below illustrates patterns and predictions of urban growth and rural decline.

Unsustainable Urban Development

The above data illustrate how urban populations have grown and are expected to grow, while rural populations progressively decline. The implications for food production and security, for maintenance of urban infrastructures, services and transport systems, for efficient safe delivery of foods to urban markets, are gravely disturbing when one recognises how little imaginative attention and investment is devoted to urban planning and management throughout the world.

These data and historical records illustrate that urban development is rarely a systematic gradual unfolding that progresses according to a predetermined plan. Unplanned and uncontrolled growth, worsening congestion, decrepit and degenerating infrastructures and services are more typical than exceptional. Urban development, ranging from at best spasmodic and inconsistent to at worst chaotic, cannot by any criteria be considered sustainable.

Urban populations

		Total (millions) Population		Annual % growth	% of total population				
	1975	1995	2000	2025	1975–2000	1975	1995	2000	2025
World	1540	2600	2900	5100	2.5	38	45	48	61
Asia	590	1200	1400	2700	3.3	25	35	39	55
Africa	104	250	305	800	4.4	25	34	38	54
Europe	445	535	550	600	0.6	67	74	76	83
Latin America	195	360	405	600	2.3	60	74	77	85

			Rural populations: Annual % growth
	1970–80	1980–90	2000–2010
Developing nations	2.1	0.9	0.3
Africa	2.0	2.0	1.4
Asia	2.1	1.0	0.7
Latin America	0.7	0.03	–0.07

Rising populations, together with increasing and diversifying demands and resource consumption, exacerbate stress on the earth's life-supporting capacity. Rising urban demands degrade available land, water, air, energy and other critical resources. Despoilment of the environment advances in scale and complexity as urban communities increase in size and density. More reliable methodologies are needed to identify interactions among demographic change, rising demand and stress on resources, services and amenities essential to expanding urban populations [WRI 1997].

Most of the world's over 20 mega-cities (cities with a population of 10 million or more) plus 300 cities each with populations in excess of one million were not designed for, nor are readily adaptable to, the population size and traffic flows now inflicted upon them.

Few of the world's large cities are coping satisfactorily with traffic congestion, atmospheric pollution, provision of safe water, hygienic sanitation and waste disposal. Such planning as is evident is frequently *ad hoc*, short term and unsystematic. Systems analysts are urgently needed to design mathematical simulation models to compare alternative systems of transportation, food distribution, delivery of health, sanitation and public transportation services, electrical and other essential energy resources.

The Confederation of British Industry estimates the accrued cost of traffic congestion in Britain in excess of £19 billion, a cost comprised of wasted working time and automobile fuel.

Energy

Antoine Lavoisier cogently commented: "La vie est une fonction chimique": life is a function of chemistry. He had sound reason: every living process depends on the conversion of one form of energy to another by biochemical reactions.

Comparative estimates of energy consumption are complicated by energy being expressed in different units by different agencies. Fossil fuels may be quoted in joules, ergs, calories, British Thermal units (BThU), kilograms or tonnes of oil equivalent, kg or tonnes of coal equivalent, barrels of oil, cubic feet or cubic metres of gas. Electricity is normally measured in kilowatt hours but may be converted to other energy units. Because the calorific value of fossil fuels varies with type, composition and source, among remote and poorer communities energy consumption cannot be accurately measured. All aggregated energy statistics cited by international agencies must be regarded as best indicative estimates. Statistical statements of available reserves are confused by variable states of accessibility and costs of extraction and processing. What price industries and consumers are willing to pay for fossil fuels influences whether the cost of extraction and delivery will be economic to fuel producers and distributors.

The following are approximate equivalents among diverse energy units.

1.0 joule = 10^6 ergs 1.0 BTU = 1054 joules

1.0 calorie = 4.19 joules 1.0 kwatt hour = 3.6×10^6 joules

1.0 kg coal = 29.3×10^6 joules =7×10^6 calories

1.0 kg oil = 41.9×10^6 joules = 10×10^6 calories

1.0 kg coal = 0.7 kg oil 1.0 barrel of oil = 0.133 tonnes.

1.0 m^3 natural gas (methane) = 38×10^6 joules

[Source: Economist Desk Companion 1992: Pub: *Economist Books* U K]

The following standard abbreviations simplify citation of the large numbers in which aggregates of energy consumption are quoted:

mega (million) = 10^6 giga (billion) = 10^9

tera (trillion) = 10^{12} peta (quadrillion) = 10^{15} exa (quintillion) = 10^{18}

Thus a gigajoule = one billion joules; an exajoule = 10^{18} joules

Table 16 quotes World Bank estimates of energy consumption by region during the 1990s, expressed in tonnes of oil equivalent (toe).

Energy Consumption 1990s

	Total 1999 Bn toe	**1989–99 % incr**	**per cap 1999 kg oe**	**Fossil**	**Nuclear 1999**	**Hydroelec Bn toe**
World	9700	+ 12.7	1600	8000	700	200
O E C D	6000	+ 10.0	4500	5000	600	130
Devg nations	3600	+ 38.5	800	2600	40	90
US/Canada	2500	+ 15.2	8100	2100	200	50
Europe	2600	–	3500	2100	300	60
Asia	2900	+ 43.1	900	2200	100	40
China	1100	+ 29.2	900	3400	900	20
Africa	275	+ 16.4	330	–	–	–
Latin America	530	+ 9.3	800	–	–	–

If all people in all nations consumed energy as extravagantly as Americans and Canadians, total world demand would rise more than five-fold and cause substantial increase in atmospheric pollution. It is hypocritical for North Americans to preach energy and other resource conservation to emerging nations until they themselves become less extravagant.

Forecast Future Demand

The World Resources Institute calculates that, during the second half of the twentieth century, world energy consumption from all sources more than tripled [WRI 2000]. A publication *International Energy Outlook* predicts that world energy demand from 2001 to 2025 will increase by some 55 per cent, at an average rate of 2.7 per cent a year. Among the fastest growing Asian economies, most notably China and India, energy consumption will double by 2025. Most energy demand will be satisfied in the foreseeable future by oil, natural gas and coal. These predictions assume little real change in fossil fuel prices, a questionable assumption given present political turmoil in the Middle East and production uncertainties among several oil and gas producers.

During 2004, assuming that energy prices would remain relatively stable, predicted annual growth rates until 2025 for natural gas were 2.2 per cent a year, for oil 1.9 per cent, for coal 1.6 per cent, and for nuclear power 0.6 per cent a year. World electricity generation is expected to rise from 13.3 tera kwh to 23.7 tera kwh, the fastest growth – more than 3.5 per cent a year – being among developing nations. Natural gas is forecast to grow from 18 to 25 per cent of fuels used in electric power generation. The assumption that prices will be relatively stable is extremely dubious, given the political disposition of the US government and the aspirations of several major oil and gas producers.

In addition to rapidly rising demands for automobile fuels, energy demand is exacerbated by cheap air travel now increasing at 7 per cent a year. It is to be hoped that, in pursuit of petroleum and diesel fuels, other nations will not follow the unethical example of the US Geological Survey, which in cooperation with US oil companies, proposes to explore and exploit oil and gas in the Arctic Ocean. If, as now seems probable, oil, gas and other fuel prices will continue to rise, non-conventional alternative energy sources could become commercially attractive.

Alternatives to Fossil Fuels

Biofuels – One might argue that all fossil fuels are biofuels since coal, oil and natural gas originated as trees and other plants that grew many millions of years ago. In contemporary vocabulary, bio-fuels are products of microbial or biochemical conversion of plant carbohydrates or lipids.

Alcohol fuels – Methanol and ethanol are the simplest alcohols used as fuels mixed with petrol for automobile internal combustion engines. Methanol may be produced from natural gas, coal, biomass from woody plants, various wastes and sewage. Ethanol results from fermentation of sugars derived by acid or enzymatic hydrolysis of complex carbohydrates, or by extraction from sugar cane. Used in conventional automobile engines, 5 to 10 per cent ethanol is the maximum that can be mixed with petrol. Higher ethanol, methanol petrol mixtures require specially designed engines, experimental in North America, and known as flexible fuel vehicles (FFVs). Alcohol fuels provide less energy per volume but emit less carbon monoxide and dioxide than petrol. In Brazil, some 60 per cent of motor vehicles use ethanol derived from sugar cane mixed with petrol.

Some years ago, in a private communication, the late Professor John Hawthorn of Strathclyde University calculated that the energy derived from ethanol combustion was less that the total energy expended in planting, cultivating, fertilising, harvesting, isolating, fermenting, distilling and distributing the ethanol produced from carbohydrates. Others have recently expressed similar reservations. Brazil justifies its ethanol fuel programme by the foreign currency saved from reduced oil imports.

Enthusiasm for ethanol derived from edible cereal starch suggests the affluent are more intent on satisfying the appetites of their automobiles than in feeding millions of hungry people in a world where arable land suitable for crop cultivation is fast declining. It would seem more sensitive to human need to derive alcohol fuels from such cellulosic waste as harvested straw, wild grasses and processed timber.

Biodiesel – During the late nineteenth century, Rudolph Diesel, born in Paris of German parents, designed an engine to compete with Nikolaus Otto's internal combustion engine that depends on ignition from a spark plug. Diesel's engine employed high compression to ignite the fuel. Originally he used powdered coal or vegetable oils under compression. Diesel's invention became widely used in many kinds of vehicle and stationary engines, powered by diesel fuel from refined petroleum. There is now interest in fuelling diesel engines with derivatives of vegetable oils.

Biodiesel fuels intended to reduce dependence on petro-diesel are mostly methyl esters of fatty acids produced from vegetable oils: in Europe mainly from sunflower, in the United States from soybeans, in East Asia from oil palm. In several Asian countries expansion of oil palm plantations to provide biodiesel results in extensive depredation of natural forests. The government of India estimates that the nation's demand for diesel fuel will rise from 60m tonnes to 100m tonnes during the next 25 years. Consequently biodiesel is perceived as an attractive alternative to, or supplement for petro-diesel. India's demand for edible vegetable oils significantly exceeds its production. Indian scientists have therefore identified several non-edible oilseeds suitable for conversion to biodiesel. Particularly attractive are inedible oils from the seeds of *Jatropha curcas*. Jatropha is of special interest, being tolerant of saline and alkaline soils, of near-arid and semi-arid ecologies. Plantations of Jatropha are being established in several Indian states on land ill suited to food crop cultivation. Indians are also exploring biodiesel derived from waste cooking oils, an example which nations addicted to fried foods would be wise to emulate.

Biogas – Produced by microbial fermentation of human and animal excreta and other waste, methane has been used as a household fuel in the People's Republic of China for many generations. A British organisation, Energy Saving Trust, estimates that methane from all human and animal waste in Britain could provide fuel sufficient for 50 per cent of the nations lorries and buses, with a consequent reduction of 95 and 80 per cent of CO_2 and NO_x emissions, respectively, from vehicle exhausts. Sweden captures and uses biogas to fuel many of its buses and refuse trucks.

Hydrogen fuel cells – Engines energised by hydrogen fuel cells emit only water vapour as exhaust. Difficulties arise from hydrogen's being plentiful but not in a free state. Electrolytic separation of hydrogen from water demands large quantities of electricity; hydrogen cannot be compressed or liquefied to equal the caloric density of petrol or diesel. Materials for hydrogen fuel cells are costly, engines that incorporate them are heavy; petrol and alcohols are more easily distributed than hydrogen. Since 1993, several governments and automobile industries have invested substantially in fuel cell research and development. Several authorities suggest that electricity generated from expanded nuclear power could economically energise hydrolytic production of hydrogen. The principle of the hydrogen fuel cell was first described in 1839, 40 years before Daimler-Benz brought forth their first petrol engine. Fuel cells are based on reverse electrolysis: by electrochemical reaction between hydrogen and oxygen. The cell, composed of two electrodes on either side of an electrolyte, is supplied with hydrogen to react with atmospheric oxygen. At the anode hydrogen atoms converted to ions release electrons as the source of electrical energy. Water is the sole byproduct. Proton exchange membrane (PEM) hydrogen fuel cells embody a solid electrolyte and operate at around 80°C. In cooperation with Ballard Power Systems, Daimler-Benz have designed and constructed a bus, big enough for 21 passengers, driven by 24 hydrogen fuel cells with power equivalent to a 160 HP diesel engine. Various vehicles driven by fuel cells are being road-tested in a number of countries, as are electricity generators of differing capacities.

If and how soon hydrogen fuel cells will significantly provide energy to vehicle engines is a matter of conjecture. Their zero carbon emissions are environmentally attractive but the cost of installation and the difficulties of distributing hydrogen to thousands of automobile service stations remain formidable constraints. Nevertheless, given the extensive interest and investment, it seems likely that vehicles powered by hydrogen fuel cells will at some future time be realised.

Some enthusiasts promote methanol fuel cells. As a liquid, methanol is easier to package and distribute than hydrogen. PEM fuel cells operated with methanol are proclaimed as promising replacements for lithium ion batteries. A recent author predicts early advent of laptop computers powered by small methanol fuel cells. Given their environmental advantages and the substantial research investments projected, PEM fuel cells will probably find economic commercial application at some time in the future.

Electric motors – There appears to be growing interest in electric motors for commercial and passenger vehicles. The main obstacle has been the low capacity of storage batteries, limiting the distance a vehicle can travel after being recharged. For short distance travel (up to 100 km) in congested urban roads, electric motors are attractive in that they use no fuel when stopped in a traffic jam and emit no exhaust. For longer journeys, hybrid engines, powered both by gasoline or diesel and electric motors that are recharged by the fossil fuel engine, are under development.

Solar and wind energy – Where sunlight is abundant, conversion of solar energy into electricity, first proposed in 1839 by the French physicist Becquerel, deserves greater attention. Where insolation intensities are high, energy unused from sunlight is equivalent to many times the world's commercial energy. Assuming solar energy conversion efficiency of 15 per cent, the total primary energy consumption of all developing countries could theoretically be provided from receptors occupying barely 0.1 per cent of their total land area, an area less than is now covered by reservoirs for hydroelectric generation. Future costs are speculative but a private communication from an experienced source predicts, between 2010 and 2025, a cost of 5c/kw hour for solar thermal and 10c/kwh for photovoltaic conversion. A recent UN report calculates that, using modern solar conversion technologies, an area of 640,000 km^2 in the Sahara could generate electricity equivalent to present total world electricity consumption.

Photovoltaic cells convert photons from solar radiations into electricity. Photons are absorbed by semi-conductors such as silicon wafers. Negatively charged electrons are released to produce a flow of electricity. Solar cells are often connected in series, encapsulated in a module covered by a sheet of glass to permit light to pass and protect the semiconductor from rain and contaminants.

Concentrated Solar Power [SCP] systems use mirrors to concentrate energy from the sun onto vessels that contain a liquid or gas that when heated to *ca* 400°C powers steam turbines. SCP has been used to generate electricity in the Californian Mojave desert for more than a decade. Some 30 countries have committed over $13 billion to a nuclear fusion reactor, an unproven and unpredictable power generation technology. A significantly smaller investment in SCP, a technology proven as effective and economic, would be more sensible and productive of clean energy and free of GHG emissions.

Windmills are an eleventh century Persian invention. During the eighteenth and nineteenth centuries, the Netherlands and Germany, respectively, drew power from 8,000 and 18,000 windmills. To generate energy equal to a modern nuclear power station, one experienced authority estimates that the most efficient modern windmills would need to cover more than 350 km^2 [Barlow 1997]. Portugal appears intent on creating a sizable wind power facility that would generate electricity sufficient to satisfy the demand of 750,000 households. Worldwide, wind power generates close to 60,000 MW. European countries appear most committed. Germany generates 18,500 MW, Spain 10,000 MW, and Denmark 3,200 MW. India reportedly produces *ca* 4,500 MW by wind-turbines.

Nuclear fission and fusion – Many nuclear reactors around the world generate electrical energy by *nuclear fission* and nuclear decay. Nuclear energy is liberated when uranium 235 (^{235}U) is so enriched and concentrated that the radioactive decay in a controlled chain reaction creates heat to produce steam to drive turbines that generate electricity. Nuclear power provides *ca* 7 per cent of total world energy, 17 per cent of the planet's electricity. Roughly 80 per cent of electricity in France, 20 per cent in the United States comes from nuclear reactors.

Nuclear power fell from favour following the 1979 Three Mile Island accident and the catastrophic disaster in 1986 at Chernobyl, when radioactive particles were carried by atmospheric currents as far away as Scotland. Contemporary renewed interest in nuclear energy results from the need to reduce CO_2emissions and the promise of safer reactors. As evidenced from Chernobyl, humans and other organisms, exposed to intense radiation, suffer severe genetic injury and fatal illness, with gross malformations in unborn living embryos. The earliest reported nuclear fission experiments were carried out in 1938 by German physicists in Berlin. The world's first nuclear power plant was constructed in Obninsk, USSR, during 1954.

Nuclear fusion generates the energy that heats the sun and illuminates many distant stars. Fusion of the nuclei of two hydrogen isotopes, deuterium and tritium, liberate vast quantities of thermal energy. Deuterium can be isolated from water; the heavier isotope, tritium, sparsely available naturally is produced by neutron bombardment of lithium. Nuclear fusion is attractive in its waste products being less hazardous than those from nuclear fission. The obstacles to the economic generation of energy by nuclear fusion are many, most important being that in all reported experiments more energy is required to cause the isotope nuclei to fuse than is generated by their fusion. The largest known experimental nuclear fusion reactor is the Jet at Oxford, England. Several European nations propose a consortium to invest at least 10 billion euros in a massive prototype nuclear fusion reactor. By 2050, it is hoped that nuclear fusion will economically satisfy much the world's energy needs. Sceptics insist that economic and efficient nuclear fusion will not be realised before the end of the century and believe the huge sum would be more productively invested in solar and wind energy.

Tidal power – There is discussion in Britain of a proposal to construct a 16 km barrier across the Severn estuary. The tides being predictable, 175 sluices would be opened to allow incoming tides to fill a huge basin in which water would be held until the tides recede, when the out-flowing water would drive some 200 electric turbines. It is claimed that the power generated would be equivalent to power from two modern nuclear power stations. The predicted cost is $27 billion. Ecologists fear the proposed installation would disrupt many wild life habitats for fish and birds.

Efficiency of Utilisation and Conservation

How much energy is consumed is influenced by how much is wasted and used inefficiently. Americans and Canadians, accustomed in earlier years to very cheap electricity and other fuels, persist in extravagant consumption by driving large motor vehicles, high power consumption in advertising signs, in public buildings and private homes, over heating during the winter and excessively cool air-conditioning in summer months. The efficiency with which motor vehicles of similar size and engine power burn fuel depends on how effectively they are maintained, how, where and at what speeds they are driven. Energy consumed to heat or air condition a building is determined by design, structure, insulation and patterns of air circulation.

Thermal energy in factory processes is wasted by steam leaking, by poorly insulated steam pipes, by process scheduling that fails to use heat generated economically and conservatively. Prudent managers conduct frequent energy audits of total consumption and use by high energy consuming processes. Large companies can install integrating watt-meters, liquid and gas flow meters to measure actual consumption. Smaller companies rely on historical and account records received from suppliers.

India is believed to lose more than 20 per cent of electric power generated during transmission over long distribution grids. Though the generation cost per unit from small-scale generators is higher than from large central installations, transmission losses from small generators located close to users are significantly lower. To supply power to small rural industries, small generators are more economic and reliable than power from a large electrical grid. Urban motor vehicles when powered by electric motors use no energy when stuck in a traffic jam and emit no obnoxious fumes. Since the mid-1930s in cities in Britain, foods have been economically and hygienically transported in electric vehicles. Electric engines operate efficiently in the higher temperatures of the tropics.

Regrettably, politicians in many nations, unwilling to offend their electorates, lack the courage and commitment to raise energy prices to levels that would encourage greater conservation. North Americans oppose a carbon tax on vehicle fuels that would serve to reduce atmospheric pollution. It is gratifying that the Canadian province of Quebec has recently imposed a carbon tax.

Atmosphere and Climate

The earth's atmosphere consists of a thin blanket of air that makes all life possible. One-half of the atmosphere lies within

5.5 km, the outer layers extend to 30 km above the earth's surface. The approximate gaseous composition of pure dry air is as shown below:

	% (per cent)		% (per cent)
Nitrogen	78	Argon	1.0
Oxygen	21	Neon	0.002
Carbon dioxide	0.035 (variable)	Helium	0.0005

Other atmospheric gases include ozone, methane, oxides of nitrogen and various pollutants [Hengeveld 1995].

The globe's climatic system may be regarded as a giant heat engine driven by incoming solar energy. Solar energy warms the earth and surrounding air, motivates atmospheric winds and ocean currents, and drives the evaporation and precipitation processes integral to the water cycle. Solar radiation is the energy source for all biospheric and photochemical reactions. The lowest atmospheric layer, which contains about 80 per cent of the total gaseous mass, is the troposphere. All energy that enters the climatic system eventually returns into space as infrared radiation.

If energy leaves at the same rate it enters, the heat engine is in balance and the earth's average surface temperature will stay relatively constant. If the energy exchange becomes unbalanced, average temperatures will change until the system adjusts to a new equilibrium. Nitrogen and oxygen are relatively transparent to radiation and exert little effect on the energy that passes through.

The remaining 1 per cent, made up of various aerosols and gases that reflect, absorb, and re-emit significant proportions of incoming and outgoing radiations, is most influential in regulating energy flows that drive climatic processes.

Climate is commonly defined as *average weather*: the average day-to-day variations in temperature, precipitation, cloud formation and winds at specific locations around the globe. Climate is a changeable phenomenon. Together with clouds, minor gases in the atmosphere provide an insulating blanket that keeps most of the earth warm, a phenomenon known as 'the greenhouse effect', caused by 'greenhouse gases' (GHG). Most important GHG are water vapour, carbon dioxide, methane, ozone and oxides of nitrogen. Theoretically, the solar energy that reaches the Planet is sufficient only to maintain an average surface temperature of –18°C but the GHG blanket raises the average surface temperature to around 15°C. The temperature difference between the tropical zones and the poles is the main driving force for atmospheric winds and ocean currents. Atmospheric moisture is transported by air currents; precipitation patterns are influenced by global atmospheric circulations. Ocean currents, topographic features and rates of evaporation influence the complex, variable precipitation patterns around the world.

The earth's natural climatic system is dynamic and constantly changing. Climate data have been measured and recorded for about 150 years. Information about weather patterns over the past one million years is derived mainly from paleoclimatic indicators. Analyses of Antarctic and Greenland ice-cores indicate a strong correlation between climatic change and atmospheric concentrations of carbon dioxide and methane, two of the main GHG. Global glaciations seem to have appeared at 100,000-year intervals, each being succeeded by a 4°C to 6°C rise in average temperature.

Temperatures in the late nineteenth and early twentieth centuries were cooler than those presently recorded. From the 1920s, average temperatures rose steadily to a peak in the 1940s, declined slightly into the mid-1970s since when warming has resumed, average temperatures reaching new highs in the 1980s and 1990s when the eleven warmest years of the past century were recorded.

The present average global temperature is 0.5°C higher than in the nineteenth century. The Planet's average temperature was the highest on record during the first four years of the twenty-first century. Despite denials by the automotive and petroleum industries and the governments they financially influence, human and industrial activities are unquestionably causing severe climate change, the consequences of which could be devastating for many regions.

GHG Emissions

It is difficult to determine precisely the quantities of carbon dioxide and methane emitted into the atmosphere, for how long they will remain, and the magnitude of their effect as GHG on climate change.

Carbon dioxide is removed by photosynthetic conversion in biotic materials, by direct absorption into oceans, surface waters and plants. CO_2 is released by plant and animal respiration, from decay of biomass, combustion of fossil and other fuels, and volcanic emissions.

Giving cause for concern are the increasing quantities of GHG released into the atmosphere, CO_2 concentrations being 20 per cent higher than at any time over the past 200,000 years. Methane concentrations have more than doubled since the late eighteenth century. Table 17 illustrates the state of GHG emissions in 1999 [Data from World Bank, WRI, Environment Canada].

It is not surprising that nations with highest *per cap* consumption of energy emit the highest *per cap* levels of GHG. Of particular concern is the rapid rise in energy consumption, in combustion of fossil fuels and resultant GHG emissions from developing nations, especially those in transition into industrialised economies.Table 17A records estimates of CO_2emissions by different nations as presented in *HDR 2005.*

Greenhouse gas emissions 1999

	CO2[Mt]	% change 1990–99	Per cap[t]	Cumulative CO2[Mt] 1800–2000	CH4 [Mt]	NOx [Mt]
World	23000	+9	3.9	1.01 M	6300	3600
Asia	6900	+38	2.1	–	2600	1200
PR China	3100	+26	2.5	73000	960	500
Japan	1200	+11	9.1	37000	60	30
Europe	5900	+10	8.1	411000	1200	600
United States	5600	+15	19.9	301000	800	500
Developing nations	8000	+37	1.8	197000	3800	2200

Mt = Million tonnes; t = tonnes

CO_2 emissions 2002

	Tonnes/cap/Yr	% World total
USA	20.1	24.4
Australia	18.3	1.5
Canada	16.5	1.9
Japan	9.4	5.2
United Kingdom	9.2	2.5
Qatar	53.1	0.2
Bahrain	30.6	0.1
Russian Federation	9.9	6.2
P. R China	2.7	12.1
India	1.2	4.7

On average, every automobile in the United States (excluding large trucks and buses) emits 20 tonnes of CO_2 every year. It is estimated that 50 per cent of all the world's exhaust emissions come from vehicles in the United States.

Human activities are disturbing the state and balance of the global carbon cycle. During the twentieth century, massive conversions of forests into agricultural land are estimated to have released over one billion tonnes of carbon. The burning of fossil fuels and other combustibles add some 23 billion tonnes of carbon dioxide to the atmosphere every year, more than 10 times the scale of emissions 100 years ago. Some 50 per cent of carbon emissions may return to the carbon cycle by absorption into oceans, terrestrial ecosystems and forests. Whether these natural carbon sinks can absorb increasing amounts of anthropogenic CO_2 emissions is doubtful given many unknowns: to what size will world populations grow, will future generations depend on present energy sources and will they use energy more conservatively? In the near term it is inevitable that nations such as China and India will increase their fossil fuel consumption with persistent increase in GHG emissions. Nor is there persuasive evidence that those who now consume most fossil fuels and emit most GHG will be converted to considerate conservationists. Without dramatic change, by the year 2050, CO_2emissions could rise to 400 times present levels.

Methane is released from organic decay, wetlands, rice paddies and the digestive processes of ruminants. Though areas of natural wetlands are decreasing, areas under rice cultivation are expanding. The numbers of ruminants that grew fourfold in the twentieth century are destined to further increase with rising demands for livestock products. Other greenhouse gases, in higher concentrations than a century ago, rise continually. Ozone, naturally present in the troposphere, has increased from various chemical processes to levels 50 per cent higher than in the nineteenth century.

Recent data indicates that since the industrial revolution of the mid-nineteenth century concentrations of GHG have risen by more than 70% and are now increasing by over 2% per year. Each tonne of carbon dioxide emitted inflicts a damage cost of about $85.

Climate and Biodiversity

Over the past 30 years, several environmental agencies in Britain have collectively studied climate change, particularly warming attributable to GHG, in relation to shifts in abundance and distribution of biological species. Predictions are derived from modelled projections of future climate change, that associate present and predicted climates including average temperatures, precipitations and seasonal variations with existing species distributions. The authors assess extinction risks among biological species covering 20 per cent of the Earth's terrestrial surface. If the lowest climate-warming projections are realised, 18 per cent could become extinct; with maximum warming projections, 35 per cent of living species could become extinct during the twenty first century. If GHG emissions are not reduced, the earth will become warmer than at any period during past 40 million years causing extensive depletion of biodiversity [Thomas *et al* 2004].

A meeting of the United Nations Environment Programme convened in Curitiba, Brazil, depressingly reports that the destruction of many organisms is occurring almost 1,000 times as fast as happened before heavy industrialisation and motor vehicles became so widespread. In its report *Global Biodiversity Outlook,* the UN lists primary causes of biodiversity degradation as habitat destruction (particularly the stripping of tropical forests), overexploitation of most natural resources, human and industrially induced climate change, all accelerating at rates that are entirely unsustainable. More than 20,000 km2 of Amazonian forests are destroyed annually by excessive, wasteful logging, by clearing to provide land on which to raise coffee and cultivate soybeans to feed animals destined to supply North American and European fast food stores.

An encouraging recent announcement by Cary Fowler, executive director of the Global Crop Diversity Trust, states: On the Norwegian island of Svalbard, a few hundred kilometres from the North Pole, a consortium of countries plans to establish a super-safe repository "for the world's most precious asset" – over one million samples of agricultural seed. Neither nuclear war, nor global warming, nor any other imaginable threat will disturb the viability of seeds stored inside aluminium packets. The GCDT, based in Rome, will manage the seed storage facility.

INTERNATIONAL CLIMATE PROGRAMMES

International organisations that study climate change, oceanic and atmospheric cycles include the Geosphere-Biosphere programme that began in 1986, the World Climate Research Programme, the Intergovernmental Panel on Climate Change (IPCC) and most recently the Global Environment Change related to Food Systems (GECAFS). The IPCC states that temperature and precipitation patterns of distribution are determined by topographic features, interactions among oceans, ice and land surfaces and atmospheric circulations. Most influential are:

a. control of GHG
b. effects of clouds and aerosols on radiation balance
c. transport and storage of heat and carbon dioxide by oceans
d. rising sea levels from polar ice melted by global warming.

Though mathematical models present different conclusions to related issues, several that simulate a doubling of carbon dioxide concentrations since the industrial revolution predict that:

a. average global surface temperatures could rise by 4.5°C, with most warming in polar regions;

b. significant changes in patterns of evaporation and precipitation, with the northern sub-tropical zones drier, and higher moisture in polar regions,

c. rising concentrations of GHG will cause climatic change, with serious consequences for ecosystems and agriculture. To avoid catastrophic climatic change nations, industries and communities must be more conservative of energy and other resources.

Scientific evidence from several reliable observers indicates that the Arctic and Antarctic are warming roughly twice as fast as the rest of the Planet and predict a rise in Arctic temperatures of 5–7°C during the twenty-first century. If present trends con[illegible] by 2050 large areas of the Arctic Ocean will be ice-fr[illegible]mmer months. Melting of Arctic ice will cause de[illegible]ng inundations of many coastal regions. In addition, more than 30 per cent of polar bears and many other Arctic and Antarctic creatures will become extinct.

Intergovernmental Panel on Climate Change [IPCC]

The Fourth Assessment Report compiled in 2006 by the IPCC is cause for serious alarm. It states:

- twelve of the past 13 years were the warmest since records were first assembled
- ocean temperatures have risen to a depth of 3 km below the surface
- snow cover and glaciers are significantly decreasing
- sea levels are rising at *ca* 2 mm/year, and
- by the end of the century, sea levels could rise by 0.5 m, deserts will spread, oceans will rise in acidity causing destruction of coral reefs, and deadly heat waves and violent storms will increase in frequency and severity.

The Stern Report

The British government Treasury Office commissioned Sir Nicholas Stern, former Chief Economist at the World Bank to report on the global economic consequences if present patterns of atmospheric pollution and global climate change persist. The Stern Report is alarming, its findings forebode devastating consequences for future life on earth if present energy generation and utilisation practices persist. At present rates of GHG emissions by 2050 the average global temperature will rise by more than 2°C. Insect-borne diseases, violent storms with flooding of coastal zones, severe regional droughts with acute water shortages will be widely exacerbated. Unless immediately arrested, GHG emissions with consequent climate change will inflict costs equivalent to 20% of global GDP. Remedial prescribed actions taken immediately would cost barely one per cent of GDP per year; resultant benefits would far outweigh the cost of inaction. Failure to act swiftly and decisively will incur far greater cost in the future and will depress global GDP by at least 5%/year.

All countries and ecologies will suffer from failure to implement immediate remedial actions. Present average atmospheric concentrations of CO_2 equivalent are about 430ppm. To realise the desirable maximum level of *ca* 450ppm by 2050 demands CO_2 equivalent reductions of at least 25%. The cost of immediate action is sustainable; failure to act quickly will impose immense hardship and excessive costs on all nations.

Contrary to the myopic pessimism of industrial and political opponents of arresting adverse climate change, creation of more conservative, less polluting energy generation technologies would yield economic benefits for existing energy industries, create new industries and expand employment opportunities in most nations.

In addition to energy from renewable resources, more efficient technologies that capture and prevent carbon emissions would permit continued, more economic use of fossil fuels including coal. All industries and communities would gain economic benefit from more conservative use and less waste of energy resources.

Specific actions proposed by the Stern Report include:

a. Encouragement with incentives for all industries and civil society to use energy more efficiently and efficiently with significantly reduced waste.

b. World-wide expansion of emissions trading. Energy-efficient nations should subsidise less-polluting and wasteful energy utilisation among emerging, industrialising nations.
c. Taxation policies that penalise polluters and reward conservationists.
d. International research cooperation is essential to devise and develop efficient, less polluting and resource conserving energy technologies.
e. Significant reduction of deforestation, particularly of natural tropical forests, and extensive programmes of reforestation. Present levels of deforestation result in more polluting emissions than are emitted by all transport vehicles.
f. Effective integration of climate change control in all governmental and international economic and industrial development policies.

Failure among all nations and international agencies to act swiftly and decisively will severely impede survival of future generations, commit the survivors to a miserable existence, and drastically destroy biodiversity by eliminating many species of organisms.

Climate Change Research

On exceptionally thorough review of recent research on climate and climate change (with 683 references) is available from the Science and Technology Branch of Environment Canada. It reviews GHG, climate during the past 400,000 years, climate model simulation and projections on how to manage the causes and consequences of inevitable climate change (Hengeveld 2006).

Water

Water is the planet's most abundant resource, covering close to 70 per cent of the globe's surface. If all the water were spread evenly across the earth it would form a layer roughly 3,000 metres deep. Almost 97 per cent of the world's water lies in oceans and inland seas, all too saline to be potable or usable for crop irrigation. Of the remaining 3 per cent that is free from salinity, over 80 per cent is sequestered in ice caps and glaciers, exists as atmospheric vapour or is buried in aquifers at various depths below ground (Hulse 1995).

In a sequence of evaporation, transportation and precipitation, water is cycled through the troposphere, each cycle lasting about 10 days. In addition to replenishing lakes, rivers, aquifers, natural and managed ecosystems, water as atmospheric vapour acts as a greenhouse gas. As clouds water contributes to climate control; rain helps atmospheric cleansing by pollutant precipitation. Levels of rainfall vary extensively across the globe, heavy rainfall being characteristic of the Amazon basin, other humid tropical ecologies and the monsoonal regions of Asia. Atmospheric, land and ocean conditions and processes interact to distribute rainfall in geographically uneven patterns, the planet's climatic zones being designated humid, semi-humid, semi-arid and arid. Arid and semi-arid ecologies suffer severe water deficits and cover some 30 per cent of South America, 33 per cent of Europe, 60 per cent of Asia, 85 per cent of Africa, as well as much of Australia and the western plains of North America. More than 80 countries, home to over 40 per cent of the world's population and 30 per cent of cropland, suffer serious periodic drought.

Drought

A discussion of what is known of the history of serious droughts, climatic conditions conducive to drought, the regions and countries where drought is most frequent was published 20 years ago [Hulse and Escott 1986]. Though meteorologists are aware of the general conditions that foreshadow drought, present mathematical models and data available are unable to predict precisely when and where serious droughts will occur. Droughts are often of extended duration, unlike hurricanes and typhoons they do not arrive and depart suddenly. Droughts of varying severity have devastated cultivated crops across the Great Plains of the USA and the Canadian Prairies roughly every 20 years since the 1850s. In the 1930s drought persisted for several years causing egregious crop devastation, soil erosion and poverty for Prairie farmers.

Geophysical evidence indicates that large areas of the African continent suffered long periods of severe drought starting *ca* 10,000 years ago. During the twentieth century serious droughts afflicted the semi-arid territories of sub-Saharan Africa every 15 to 20 years. Particularly disturbing is evidence from West African weather stations that suggest a persistent downward trend in annual precipitations since the 1950s. To what extent drought in Africa will be exacerbated by global warming, aggravated by increasing emissions of GHG, is an issue of contentious debate. Weather conditions over sub-Saharan Africa are influenced by prevailing atmospheric circulation systems, especially interaction between the sub-tropical high-pressure zone and the intertropical convergence zone (ICTZ).

As determined by GNP/cap, 20 African countries that are most at risk from recurrent drought are among the world's poorest. For farmers in Australia and North America, some protection is provided by crop insurance and government interventions at times of severe drought devastation. Such programmes of protection are unavailable to most poor African farmers. The total cost to human life and health of African droughts is incalculable. While the plight of thousands with strength sufficient to walk to refugee camps when drought strikes is made known by the world's television and media, many more die unnoticed in their homes or while migrating in search of food and water. Nutritional deficiencies prevail in months immediately before harvest when grain stocks are lowest. Protein-energy malnutrition, anaemia, diarrhoeal diseases, hepatitis and measles, common at all times, are aggravated by drought-induced water deficiencies.

Drought and Crops

Cereals and legumes are hygrophytes and need water during their growth cycle. Drought soon after seeding restricts germination; at time of germination drought can desiccate the plantlets before they flower and fruit. Floral development and pollination are highly sensitive to drought-stress which results in a reduced number of seeds. During seed filling drought results in small, shrivelled seeds because of reduced photosynthesis and translocation of carbohydrates from leaves to seed endosperms. Low soil moisture inhibits uptake of soil nutrients through plant root hairs. Plants debilitated by drought are susceptible to infection, infestation, invasion by competitive drought-tolerant weeds, and attack by predators. Ravages by locusts following prolonged drought are familiar throughout the African drylands. During the Sahelien drought of the 1970s, sorghum in northern Senegal suffered severe attacks by wild rodents deprived of their natural food sources.

It is sometimes suggested that drought tolerance could be enhanced by genetic modifications. But different plant species use diverse mechanisms to resist drought. Specific adverse effects of drought differ at different stages of growth and development. To imbue a plant with resistance to drought at all stages of growth would be genetically complex. Over the foreseeable future, drought inevitable and unpredictable will inflict severe hardship wherever it strikes.

Water: Available and Withdrawn

Table below, with global and regional data drawn mainly from WRI reports, quotes estimates of available water resources, and water withdrawn in 1995 for various uses.

Water Resources 1995

	Available		Withdrawn		Withdrawal %		
	Total (km^3)	Per cap (m^3)	Total (km^3)	Per cap (m^3)	Industry	Domestic	Agriculture
World	41000	7200	3300	645	8	23	69
Africa	4000	5500	145	200	7	5	88
Asia	13200	3800	1630	540	6	9	85
L America	17300	21000	195	410	12	34	54
Europe	6200	8600	455	630	14	55	31
U S A	2500	9400	470	1900	13	45	42

Competition for Scarce Water

Roughly 70 per cent of the Planet's water and more than 80 per cent in developing countries is used in agriculture, mostly for crop irrigation. Competition for water from industry and purposes other than crop irrigation are progressively increasing. Total world water withdrawn is predicted to rise from *ca* 3,300 km^3 in 1995 to *ca* 4,800 km^3 in 2025, a 50 per cent increase. Over the same period, water for non-irrigation purposes is expected to rise by over 60 per cent, for domestic use by more than 70 per cent, greatest increases being in nations now or once described as 'developing' [Rosegrant *et al* 2002].

By 2025, industrial water consumption will be higher in developing than industrialised nations. Less water available for irrigation forebodes that crop yields among developing nations, now increasing on average by 1.9 per cent a year, will fall to barely 1.2 per cent a year, equivalent to an annual loss of 130 Mt of cereals. While water for irrigation and crop production are adversely affected, world demand for cereals is projected to grow by almost 50 per cent, by some 800 million tonnes. A likely distressing consequence will be the need for poorer nations to raise their food imports from 105 to 240 million tonnes per year by 2025. Most serious shortfall will be in sub-Saharan Africa where imports of cereals are forecast to triple [IFPRI 2002].

Waste and Mismanagement

Urgently needed is a worldwide improvement in water use efficiency, in particular for irrigation. Poor irrigation is wasteful of water, depletes groundwater resources and is the cause of salinity and loss of nutrients in soils. To use water more conservatively and efficiently demands technological innovation and administrative reorganisation. In a nation familiar to this author, from a reservoir constructed for hydroelectric power generation, water is gradually released and channelled over long distances for crop irrigation. Between the reservoir and distant farmers' fields the water passes through nine different ministries and agencies of national and state governments. It is not surprising that much water is lost along the way since no single body bears overall responsibility and authority for management and control. When rivers flow through several national territories, administrative difficulties compound and disputes over water rights can lead to violence. Some political analysts anticipate military conflict between Middle Eastern nations who compete for access to limited water supplies. Wasteful misuse is apparent among people who migrate from places with abundant water to arid ecologies. Excessive and extravagant water consumption by foreigners who have taken up residence in new settlements in a Middle Eastern country are deplored by neighbours born in the region and accustomed to use water prudently.

Unlike farmland and buildings, inland lakes and rivers are regarded as common property, free to be plundered or polluted by anyone so disposed. As the worldwide demand for a scarce and precious resource rises, aquifers are being wastefully depleted, surface waters polluted by domestic and municipal effluents, by agricultural and industrial chemical contaminants. Demands for policies that integrate economic with ecological considerations appear to fall on deaf ears.

Future Prospects

A contemporary review of world water prospects is published by the International Food Policy Research Institute [Rosegrant *et al* 2002]. The review concludes that the water shortage crisis, predicated on an anticipated worsening of several critical factors, forebodes a breakdown in water supplies for hundred of millions of people, to devastation of wetlands, serious reductions in food production with consequent rise in food prices. Excessive diversions of water flows and overdrafts of ground water will lead to water shortage and related environmental degradations in many parts of the world. In the absence of policies that encourage protection and conservation, competition for water among domestic, farming and industrial consumers will inexorably increase. The review recommends several courses of action:

- investment in infrastructures to increase supply of water for all essential purposes;
- enactment of policies to conserve and enhance efficiency of use;
- investment in research and development designed to raise crop productivity per unit of water and land.

In relation to the last recommendation, it would seem wise, where water is in scarce supply, to encourage farmers to cultivate crops that provide most economic returns to water. In some semi-arid African nations, farmers have switched from subsistence grains to high-value horticultural and floricultural species for export or sale to more affluent communities. Farmers who receive adequate compensation for their produce can be charged realistic prices for the water they use.

Tim Flannery, an Australian climatologist, paints a grim picture of the Planet's water resources. Many of the world's great rivers – the Indus in Pakistan and northern India, the Yellow River in China, the Colorado in the United States and Mexico – are being seriously and irreversibly depleted. Over-draughts of many fossil water sources, accumulated over countless ages, exceed their annual rates of replenishment by aqueous precipitations. The worsening world water crisis is aggravated by climate change and by pollution of inland surface waters. Future water shortages will be most severe in drought-prone regions where already many millions do not have access to adequate safe potable water [Flannery 2006].

A World Bank Perspective

In a lengthy policy paper, the World Bank reports how its investments in irrigation, water supply, sanitation and hydro power have encountered operational and social difficulties, water of poor quality for which consumers were unwilling to pay, with poorly managed investments in measures to increase water supply and improve conservation. It admits that the Bank and governments have given too little attention to environmental considerations.

The report cites four specific difficulties: a. fragmented public policies and investment programmes; b. failure to recognise interdependence among involved agencies and users; c. neglect of economic pricing and sound financial accounting; d. too little attention to water quality, public health and environmental protection. In its 1993 report the Bank promises future activities will be designed to benefit industries, agriculture, power generation and consumers with cost-control measures commensurate with financial viability.

Water used by Industrial Biotechnologies

Food processors take in only 4 per cent of the total water consumed by Canadian industries. Nonetheless, all food processors use sizable quantities of clean water: as an ingredient of many foods, as steam, as ice, and for cleaning of raw materials, employees and their clothing, equipment and factory structures, from all of which much waste water emerges. Though governments enact laws to control the quality of industrial effluents, enlightened industrial managers install means to purify and recycle wastewater; and to remove obnoxious or offensive pollutants from all wastewater released [Amos 1997]. Pharmaceutical industries must remove all substances that pose potential hazards to health or ecosystems if released into surface waters and public sewage. Biotechnological industries should purify wastewater to be usable at least for factory cleaning or conversion to steam [Jogdand 1996].

Waste Water Treatment

Impurities in waste water are categorised as: a. Suspended solids (SS), b. Biological Oxygen Demand (BOD5), c. Chemical Oxygen Demand (COD); levels of acidity or alkalinity may also be determined. SS covers solid particles large enough to be retained in filters of prescribed porosity. BOD5 estimates oxygen uptake over 5 days at 20°C to oxidise all biodegradable organic matter. COD determines the quantity of oxygen to oxidise all organic and susceptible minerals. The following table lists typical SS, BOD and COD values for different effluents [Wheatley 1994]:

	BOD (mg/litre)		COD (mg/l)	SS (mg/l)
Domestic sewage	300–600	Meat, poultry	500–8500	75–4500
Silage effluent	30,000+	Dairy	80–9500	25–4500
Vegetable washing	1600–11,000	Fruits, vegetables	1600–1800	Variable

Jogdand cites BOD values for pharmaceutical, pulp and paper industries' effluents together with a review of biotechnological methods of remediation and purification [Jogdand 1996, 2003]

	BOD (mg/l)		BOD (mg/l)
Sulphite liquor/		Antibiotic	
paper mills	20000–45000	waste	5000–30000
Distillery	10000–25000	Antibiotic	
stillage		spent liquor	2500–7000

Waste water treatment proceeds in three steps [Wheatley 1994]:

Primary: Physical removal of solids by sedimentation, filtration or activated sludge;

Secondary: Biological degradation of organic matter by aerobic and/or anaerobic microbial fermentation;

Tertiary: Biological and/or chemical inactivation of toxic, organic and inorganic materials resistant to (1) and (2).

Final Comment

As it relates to development, 'environment' is virtually impossible to define within precise parameters. As the foregoing illustrates, environmental resources are immensely diverse and difficult to assess accurately. The global environment is complex beyond human comprehension. Recommendations anent global environmental conservation can be prescribed only in broad generalisations since conditions, resources, opportunities and constraints vary extensively among regions, ecologies and communities. There being no international organisation with authority and power to enact and enforce regulations deemed desirable to protect and provide for future generations, recommendations can be little more than hoped for good intentions. Among governments, organisations and individuals, opinions differ widely on desirable courses of action, some differences being honest and intellectual, others motivated by egocentric avarice with no wish to change a wasteful way of life.

It would appear that to many automobile producers, the politicians and governments they support, influence and control, motor vehicles and their appetites are of far higher priority than the needs of millions of impoverished humanity for food, clean water and healthy lives. A statement by a recently retired senior executive of one of the world's largest petroleum corporations, while he received millions of dollars in retirement benefits, that there is no scientific evidence for global warming caused by GHG, was repugnantly greedy and utterly irresponsible.

More than a decade of contentious negotiations resulted, as of April 2006, in 163 countries signing the Kyoto accord that proposes substantial reductions in atmospheric pollution. It is encouraging that, despite the refusal of the US Republican government to ratify the Kyoto accord, the State of California and close to 200 American cities, with combined populations in excess of 100 million, have adopted climate conservation protocols similar to Kyoto accord prescriptions.

Among persons and agencies with genuine serious concerns there seems broad agreement on the need to protect the atmosphere from pollution and human activities conducive to climatic change; to protect fragile ecosystems and conserve biodiversity. The enlightened agree also on the need for informe planning of land use, urban growth, water and energy utilisatio and conservation; for equitable opportunity in internation trade and access to essential resources; and for the affluent be more generous in support of the less fortunate.

That being said, these are all issues remarked upon by th Pearson, Brandt and Brundtland Commissions, by countless subsequent international conferences and working groups. The tangible responses to these issues have been far from what the originators hoped for. What should be specific objectives for sustainable environmental development remain contentious and elusive and will probably remain so during the immediate future. The greatest hope lies in exemplary policies and actions by particular governments and communities, policies and actions demonstrably beneficial that others could adapt according to their needs, opportunities and resources.

The costs, insults and injuries to the healthy lives of future generations caused by present extravagant use and waste of energy, water and other essential resources, together with pollution of the atmosphere, cannot be predicted. Many serious observers, most recently Sir Nicholas Stern, believe the consequences will be catastrophic. It is therefore incumbent on all governments, societies and industries to assume the worst, to adopt practices and life styles more conservative and protective of the environment and the planet's resources.

Little will be gained by convening more international conferences. Urgently needed is an international organisation recognised by all nations as having the authority to prescribe standards and regulations for biodiversity and natural resource conservation, for atmospheric protection, to restrict severely and reverse global warming caused by human and industrial activity. Since such regulatory control would demand a less wasteful and politically unpopular life-style, it is unlikely to be proposed or adopted in the foreseeable future. Regrettably, the insatiable greed of the most affluent and powerful nations does not inspire confidence that Brundtland's objective of safeguarding the needs of future generations will come to fruition.

Marie Antoinette is reported to have said: "Let them eat cake". Contemporary affluent and powerful politicians and those who sustain them seem to say: "We can be as wasteful as we please; let future generations suffer and, if need be, suffer death by starvation".

Al Gore, a former US Vice-president, in a pamphlet accompanying his film *An Inconvenient Truth*, lists "ten things to do to help stop global warming". Among them are the following: drive less (every two miles you don't drive saves 1 kg of CO_2); recycle waste; use less hot water; plant trees (a single tree will absorb one tonne of CO_2over its lifetime); avoid products with a lot of packaging (you can save some 500 kg of CO_2a year if you cut down your garbage by 10 per cent). These are simple things to do, if the affluent would only take heed.

It is clearly evident, given the political dispositions of American, other affluent governments and the industries that support and encourage them in their wasteful depletion of scarce resources, now evident among emerging industrialising nations, that there is little governmental concern for the needs of future generations, many of whom with economic and environmental conditions worsened by global climate change will be condemned to lives of extreme poverty, hardship and misery. The US and Canadian governments, who until recently refused to admit that global warming caused by GHG emissions was a scientific fact, now reluctantly claim to be partial believers. It is too soon to know whether they will enact legislation that will offend their industrial supporters in the fuel oil and automotive industries. In any event anything less than drastic action will be too late to reverse the damage done and inevitable future devastations.

CASE STUDIES OF SUCCESSFUL PROJECTS

In the previous ten chapters the author has taken a critical look at what is often termed 'sustainable development' and has argued continuously for a clear definition of terms of both those words, not only as concepts but also in their application to many different circumstances, from food production and food security to industrial biotechnologies. Too often terms are ill-defined and consequently the objectives are open to misunderstanding. Too often the research is focused on a first, part-way objective – for example, the breeding of higher yielding rice genotypes – and the attainment of the real objective, the adoption of new technologies by small farmers, is either unplanned or else taken for granted. This book has laid heavy emphasis on the importance of a systems approach that rejects any top-down practices and involves the intelligent participation of the beneficiaries from the outset.

This chapter moves from setting out such argumentation to offering a number of examples of where development has been planned on a systems basis, and where the outcome has been sustainable and beneficial. The examples are drawn from Latin America, Africa and Asia, and the International Development Research Centre has been supportive in all but one of them. The exception is the Canada-Mysore Project, where men and women from 45 nations have had training in food technologies; it was launched a decade before IDRC's own birth. The first example of IDRC initiatives concerns its advocacy of farming systems research, to take results from the experimental plots of the international (or national) research institutes to sustainable application in ordinary farmers' fields.

Closely connected is the work done in the creation of a new generation, beyond IRRI and CIMMYT, of international agricultural research institutes: ICRISAT and ICARDA, to focus on research in the semi-arid tropics and other dry areas. With the same motivation, of applying a systems approach to forestry for the purpose of lasting human benefit and resource management, IDRC first launched a social forestry programme and later helped establish ICRAF, the International Centre for Research in Agroforestry, in Nairobi. There is a similar thread of resource management in the fifth case-study, "Local management of water", in which management of demand is particularly stressed, as well as (significantly in China) the involving of the local people affected in selecting the technical solutions.

Once again, the systems approach, starting with a thorough diagnosis of a particular situation by a trained local team, is applied in the case of "Essential Health in Tanzania". And finally, the example in the Pondicherry region of linking a rural population by radio and computer to a wide range of relevant information is an indicator of how imagination and targeted aid with some modern technologies can transform communities.

THE CANADA-MYSORE PROJECT

In 1960, the British government convened a conference in London to mark the centenary of the world's first food and drug act, enacted in 1860. During the conference, Joseph Hulse, then research director of one of Canada's largest agribusiness companies, was asked by an old friend, Sir Norman Wright, who was deputy director-general of the UN Food and Agriculture Organisation (FAO), if Canadians would be willing to support FAOs recently launched Freedom From Hunger Campaign. Specifically, Wright asked for financial support for an International Food Technology Training Centre. The Indian government had promised physical facilities for such a centre in a former palace of the Maharajah of Mysore.

An FAO study had revealed that no university, college or other institution anywhere in Asia was providing training in practical means of food preservation and its safe distribution.

Countless tonnes of perishable foods were being spoiled and wasted after harvest. Money was needed to cover travel, accommodation and related administrative costs of bringing men and women to the Indian government's Central Food Technology Research Institute (CFTRI) in Mysore, southern India. CFTRI employed a staff of scientists, most of them with degrees in natural and biological sciences from foreign universities. These Indians, together with invited lecturers, would provide training in food preservation technologies to Indians and other Asians.

Early in 1960, Canada had established a small Freedom From Hunger Campaign (FFHC) committee chaired by Mitchell Sharp, a former senior public servant who later served as Foreign Minister in the Liberal government of Pierre Trudeau. The FFHC committee agreed to undertake what became known as the Canada-Mysore Project as its principal contribution to the FAO campaign and appointed Hulse as the chair and director of the project.

During a visit to India, Hulse took and collected many photographs illustrative of vast quantities of harvested crops and livestock products that were degraded and spoiled for want of effective preservation. Adrienne Clarkson, then a CBC producer, was the first journalist to interview him on this project. Subsequently, sets of photographic slides with accompanying explanatory notes were provided to Canadian scientists and others, who gave talks to service clubs, schools, religious and other organisations to gather support for the project.

Relatively soon, the presidents of Canada's 20 largest agribusiness corporations formed an industrial fundraising committee, while the United Nations' Association in Canada, Oxfam, Save the Children Fund and several other NGOs agreed to support the project. The prime minister and the leaders of all parliamentary parties became official patrons; the army, air and sea cadets and many high schools raised money by selling ballpoint pens "To write-off hunger". Local fundraising committees established across the country gave valuable support.

The training programme at Mysore consisted of a series of short courses and a longer, two-year course. The first short course, organised by two Canadians, dealt with storage and de-infestation of cereal and legume grains. During a later visit, Hulse obtained the consent of the University of Mysore to award M.Sc. degrees to all who successfully completed the two-year course. Every major agribusiness company in Canada provided the money for one or more M.Sc. students, as did each of the four Canadian Banks. This was the first time that Canadian banks and most industries had contributed to a charitable project outside of Canada. Canada's two most loved entertainers, John Wayne and Frank Shuster, became honorary co-chairs and recorded 30-second advertising spots that were broadcast across Canada.

The fundraising began in 1961. By 1964 the first short courses began; in 1965 the first batch of 20 M.Sc. students arrived, each from a different Asian nation. Monies raised in Canada supported the IFTTC students during the centre's first seven years of activity. Later, other nations contributed, the Swiss began a school for milling and baking and, in the 1970s, IFTTC became a campus of the United Nations University.

When the project began, India was receiving substantial quantities of Canadian wheat and other food resources in the form of food aid. From the sale of this food, the Indians acquired resources that were invested to build a hostel where the visiting students were housed and fed.

The training programme still goes on. Since it started, more than 7,500 men and women from some 45 developing nations have taken courses at Mysore. Though most were Asians, students from Africa, the Middle East and a few from Latin America have benefited from various training programmes. They have come from as far abroad as Mexico, the Sudan and Zambia. Several subsequently advanced to become directors of food research in Korea, Singapore, Sri Lanka, India and some other nations. Other Asian nations have since created relevant training facilities, several with staff trained at Mysore.

The IFTTC is now self-supporting with its own administrative staff. The courses are so popular that there are

20 to 30 applicants for every place in the two-year M.Sc. course. More than 90 per cent of the graduates are employed in industry. It might be argued that India has benefited most from the Canada-Mysore project, but trainees and graduates from many other nations have returned to teach or work in food industries. Illustrative of the overall benefit: food processing and distributing are the fastest growing industries in several Asian nations, not least India where in 2005 the value of industrially processed foods was over 1,000 times the value when the Canada-Mysore Project started in 1961.

The project has been successful because:

a. the objectives, resources available and needed were clearly defined;
b. Canadians created an effective organisation to support and monitor the project;
c. the Indian government, the scientists and technologists at Mysore have maintained a sustained dedication;
d. the Indian food and agribusiness industries have recognised the value of the men and women trained and employed them, and have provided financial and technical support for the Centre;
e. Hulse, in continuing as Canadian Chair and Director of the Canada-Mysore Project, has maintained an active interest in the Mysore Centre and its progress over the years, and during the past 12 years has acted as Visiting Professor, lecturing in agribusiness management and food systems analysis and integration.

Farming Systems Research – IDRC Initiatives

The IDRC Board set the Centre's agricultural staff a special challenge when it determined to place high priority on marginal and impoverished lands and ecologies. For ecologies (then and now) differed widely, as did land use and water management; and the basic infrastructure for these lands and access to markets were scarcely developed. Scientists had done little work on how small farmers managed to eke out a livelihood from any particular quality of land. There was, in fact, in 1970 very weak scientific underpinning on which to base any development help to farmers. Three decades have seen great change with the worldwide acceptance of farming systems research (FSR). And much of the change and the evolution of FSR can be traced to IDRC's early initiative and consistent support.

Latin America was a logical region in which to start, since researchers in Colombia and Mexico had turned to studying the realities of small-scale farmers who were interplanting crops of corn, beans and potatoes. But about the same time, IDRC president David Hopper, who had worked with the Rockefeller Foundation in India, learnt that an American soils scientist had begun a 'farming systems' project at the International Rice Research Institute (IRRI) in the Philippines, and asked Joseph Hulse, director of IDRC's Agriculture Food and Nutrition Sciences division (AFNS) to see if it warranted the Centre's support.

Hulse found that the project was largely confined to a soil and water study related to rice cultivation, but IDRC agreed to help by financing two soil scientists to produce a soils map of Asian rice-producing territories. As well, through former colleagues in the FAO Land and Water Division, Hulse learnt that FAO had an immense collection of water resource studies in Asia, which had never been systematically collated and analysed. He recruited a post-doctoral fellow, who spent weeks at FAO sifting through the studies and compiled a summary report that was given to IRRI.

Early in the 1970s the Rockefeller Foundation transferred the agronomist Richard Harwood to IRRI to take charge of FSR, but he soon found that the IRRI Board was opposed to any activity other than the breeding of higher-yielding rice genotypes. Fortuitously, its Director-General resigned, and the Board appointed Nyle Brady, Dean of Agriculture at Cornell University, as the next Director General.

Hulse seized the opportunity and visited Brady at Cornell on three occasions to urge that IRRI include a significant FSR component in its programme. He drew the analogy with marketing research (having been Director of Research at the Canadian agribusiness Maple Leaf Mills). Before launching a new product, food companies identify specific consumers and consult with them about the quality, quantity and price likely to be accepted. Hulse suggested that IRRI researchers likewise determine the capacity of Asian rice farmers to adapt to and adopt new rice production technologies. Brady expressed serious interest, and Hulse promised to seek IDRC support for a broader programme.

The Caqueza Project

Meanwhile, IDRC had decided to support the work of the Colombian Farming Institute (ICA), which was starting to study the production methods of farmers in the mountainous region of Eastern Cundimarca. The focus was on nine municipalities centred on Caqueza, and IDRC contributed the skills of Dutch-born agronomist Hubert Zandstra.

Much of the early ICA work there was demonstrating improved methods, rather than researching the local realities and diagnosing the constraints. But it shifted during Zandstra's four years to addressing the factors that may influence farmers in (or deter them from) adopting new methods. He listed them in the book *Caqueza: Living rural development* he co-authored: risk, credit (perhaps rotating funds), marketing, training and buffer institutions (including time for reflection). One main lesson from Caqueza was that the small-scale farmers must have meaningful participation in the research, including sharing the setting of the agenda.

In 1973 another IDRC scientist Gordon Banta visited Caqueza from his base at IRRI in the Philippines. He had been sent to IRRI by David Hopper as a social scientist to help broaden their research for small farmers beyond work on improving crop yields on IRRI's own land. Banta had worked with Richard Harwood on diagnostic studies and field trials on farmers' fields in Batangas province. Banta carried back the Caqueza experience, and in 1974 he and Harwood organised a meeting at which Asian, Latin American and African researchers shared and synthesised their experiences of the past decade. Out of this meeting grew the Asian Cropping Systems Network, which IDRC supported for many years.

Until this network began to influence attitudes, many national institutions did not allow their researchers to conduct their work in farmers' fields. Research managers saw little need to involve, from the outset of introducing new technologies, the farmers affected by it. Some indeed believed that soil and environmental conditions faced by small-scale farmers were similar to those on experimental farms. It took years to change these top-down attitudes in some countries. The on-farm research in the 1970s involved some farmer participation, but there was little methodology developed.

Ilo-Ilo and Diversifying FSR

When Harwood resigned to return to the United States, Zandstra moved to IRRI for five years and helped face this lack, by leading an expanded and diversified FSR programme. Much experimenting with methods of systems research took place. A major innovation was to consider as variable factors in land use not only such matters as soil, weather and economic and social conditions, but actually consider land use itself a variable. Did land that had always been planted to rain-fed rice always have to be so? With a 90-day growing season for new high-yield strains, a farmer could also grow cowpeas or soybeans or a second rice crop on the same land – or maybe grow some cash crop and no rice at all. It took time to persuade top staff at IRRI that rice was not a "given" in this method, or those at CIMMYT who were moving into African research that maize was not. IDRC supported experiments of this sort at three sites in the Philippines including a base in Ilo-Ilo, a distance away from IRRI at Los Banos.

Farming systems researchers explored in several other directions. The FSR programmes combined concepts differently and produced many variations in types of farmer participation, the importance placed on diagnostic methods, the market-based analyses, and the dynamics of rural families, which began to give importance to gender analysis. Work done at Ilo-Ilo alongside IRRI produced not only a general broadening of minds about land use; it also led to the formation of a group of female professionals who brought gender aspects to the forefront. It also supported in various countries the establishment of Farmer Field Schools and local agricultural research committees, bringing in not simply farmers but also people at various points along the food and market chain.

On-farm research in land use during the 1980s helped increase productivity in many crops – corn and potatoes, legumes and rice, and also crops and livestock. It is fair to claim that these developments would not have happened if research had only been carried out in single-crop studies or individual cattle-farming enterprises. The FSR approach became adopted by most national centres, as well as in newly created international centres in agroforestry and livestock research (both based in Africa) as part of their official programmes.

The Brundtland report and the Rio conference's Agenda 21 added the challenge of taking into fuller account in the FSR approach the environmental aspects and the sustaining of natural resources. Researchers now had to measure existing and alternative resource uses against these factors and deal with complex trade-offs in land use.

Zandstra, who in the 1990s moved on to head the international potato centre (CIP) in Lima, Peru, made it a priority to build the capacity to manage the natural resources in an unquestionably sustainable way, and in this FSR was an ally. He says: "It is evident and encouraging that FSR has evolved to a set of effective participatory research approaches". And he gives credit to the Ottawa centre: "It is still easy to recognise the impact IDRC has had on the scientists and their institutions through three decades of support for FSR. It has done so by being consistent and by finding important allies in the donor community".

Comment from M.S. Swaminathan

Professor M.S. Swaminathan, former Director General of the International Rice Research Institute, makes the following comments on the subject of the Farming Systems project:

IDRC's visionary programme of Farming Systems Research opened up new opportunities for work, income and ecological security for rice farming families in Asia. The IDRC support of FSR led to a synergy between ecology and economics, and provided work and income security throughout the year. It helped to maximise the benefits of cubic volume of air, soil and water. It showed how we can link paddy and prosperity together by adopting appropriate crop and livestock in efficient integrated farming systems. Farming systems research spearheaded by IDRC helps to achieve a paradigm shift from the Green [Revolution] to an Ever-Green Revolution designed to enhance productivity in perpetuity without associated ecological harm. During the period I was Director General of IRRI, we observed that the farming systems research led to the integration of efficient stem modulating nitrogen-fixing species. Also FSR helped to develop the international fertiliser response network into an international sustainable rice farming network.

During the twenty-first anniversary of IRRI in 1985, outstanding farming women and men who participated in a special discussion all mentioned that the FSR has opened a new chapter in the art and science of rice cultivation. It was therefore an act of great vision on the part of Dr J.H. Hulse and IDRC who initiated these important research programmes far in advance of the [general] realisation that rice cultivation will have a future only if the principles of ecology, economics, gender and social equity, and income generation are integrated in the form of a farming systems research programme.

Employment in Rural Agribusiness

Rural people can be liberated from poverty when enabled to earn a stable income from sustained employment in agribusiness enterprises which, individually or collectively, they own and operate. Agribusiness includes production of crops and livestock; of materials essential to crop and livestock production; preservation and processing of agricultural, horticultural and livestock products; and safe delivery to identified markets. Sustained development of rural agribusiness requires:

a. a comprehensive, critical assessment of resources available and resources needed to produce and deliver familiar, acceptable, affordable products to identified accessible markets;

b. training to acquire relevant practical knowledge and skills;
c. constant, consistent, reliable access to technical, financial and marketing advisory services;
d. organisation into legally registered community cooperative associations administered by the community. A community cooperative organisation can exert greater influence than isolated individuals in gaining favourable access to loans, credit, bulk purchasing and markets; in providing essential supporting and advisory services;
e. two highly successful programmes in south India are exemplary.

The MSSRF Biovillages Programme

In 1991, the M. S. Swaminathan Research Foundation (MSSRF), based in Chennai, began its Biovillages programme for rural agribusiness development. The programme persuasively illustrates that, given the opportunity, with effective training and advisory services, poor rural women with little formal education can collectively and individually engage in sustainable agribusiness. MSSRF supports a national biovillage network, illustrated by the programme in Pondicherry. The Pondicherry programme is administered by a Biovillage Council of some 30 women representatives from 19 rural communities. A Biocentre is staffed by young trained advisers recruited from the biovillage communities who train, assist and provide an immediately accessible advisory service to the rural women, negotiating low interest loans and credit, providing veterinary and ancillary advisory services. Each village community is equipped with a computer linked to market outlets to provide instant access to current prices and levels of demand for products offered by the biovillages. Computer graphics and illustrations are useful tools for demonstration and training.

Pondicherry biovillage women cultivate horticultural crops, produce low-cost hybrid seed for more than a dozen vegetable species, organic fertilisers from vermiform compost, and bio-pesticides from Neem trees and other local resources. The women raise poultry, culture fresh water carp in village ponds, compound fodder for their cows. The Biovillage Centre maintains demonstration plots to ensure efficient crop cultivation and crop combinations that contain most nutrients essential to human health. The Biovillage Women's Council study opportunities for new and expanded activities, manage and monitor collective financial resources, and make bulk purchases at wholesale prices of essential supplies.

VGKK/KT-CFTRI Health and Employment Programme

The Vivekananda Girijana Kalyana Kendra/Karuna Trust is an NGO registered in India that provides health care, basic education, vocational training, biodiversity conservation, stimulates community self-help and small-scale rural industries in the B. R. Hills/Yelandur taluks of Chamarajanagar district of Karnataka State. The forest reserve covers more than 350 square kilometres and is inhabited by some 80,000 people of the Soligas tribe whose ancestors have subsisted on the products of the forest for countless generations. Several years ago an eminent Indian physician abandoned his practice in Bangalore to provide much needed health care to the poor tribal forest people. At the outset, more than half the children suffered chronic malnutrition; infant mortality exceeded 150 per 1000 live births; leprosy, epilepsy, tuberculosis and diarrhoeal diseases were hyper-endemic. Assisted by other volunteer physicians and surgeons, the director provided a diversity of remedial and health care services. They established clinics for maternity, ante-and post-natal care, a surgery, ward residence facilities, a mobile surgery to visit remote communities, and trained young tribal women as nurses, midwives and medical assistants. By multi-antibiotic drugs and reconstructive surgery, the incidence of leprosy dropped from over 20 per cent to less than one per cent of the population. Comparable improvements occurred with other hyper-endemic diseases.

The need was evident for education, vocational training and employment to alleviate abject chronic poverty. Schools for primary, secondary education and vocational training were established. The physician director then requested assistance to promote rural agribusiness industries to harvest and process non-timber forest products (NTFP) and horticultural crops cultivated in forest clearings. Extracts of more than 60 identified natural medicinal plants have been registered as Ayurvedic drugs. Opportunities by which to expand from domestic to small commercial scale processing of fruits, vegetables, wild honey and medicinal plants from the forest were realised through a grant from a development agency and a loan from the local agricultural bank. Roughly 50 poor women were trained to operate and manage factory-scale processing technologies by which to transform various NTFP into marketable products, products familiar and readily acceptable to local people. The factory, with a working area of 220 square metres, is equipped for steam processing, fluid concentration, dehydration and simple fermentations. An ingenious evaporative cooling chamber preserves fresh and perishable biological materials.

The factory, constructed entirely of local materials by local labour, was completed in 45 days and began full-scale production in September 2004. Technologists from the Indian Central Food Technological Research Institute (CFTRI) at Mysore trained the women, gave guidance in product design, development and quality control, and continue as an accessible, reliable source of technical and marketing advice. The VGKK organisation is formally registered as a cooperative association and administers a credit union from which members can obtain loans or make interest-bearing deposits. In addition to the 50 women who manage and operate the factory, cultivation and harvesting of raw materials, marketing of processed products, provides employment for very many more. Among the tribal people, it is a self-imposed discipline that quantities of NTFP harvested never exceed the natural rate of regeneration. Thus it is a model of biodiversity conservation.

These two examples clearly illustrate that poor rural people who have little formal education can be trained to produce and process a wide diversity of biological materials, so long as they have constant and consistent access to technical, financial and marketing advisory services, services provided by intelligent young people indigenous to and therefore familiar with the language, culture and customs of the people to be assisted. They illustrate that rural poverty alleviation demands very many relatively small financial investments; that poverty alleviation is a labour-intensive not a capital-intensive undertaking, a lesson yet to be learned by many development agencies.

Focus on Semi-arid Tropics

During its first meeting in October 1970, the IDRC Board of Governors recommended that the Centre's programme give priority to rural poor. Joseph Hulse, then the director of the Agriculture, Food and Nutrition Sciences division (AFNS) proposed, given the Centre's limited resources and the many millions of rural poor, that AFNS concentrate on the poorest of the poor in the semi-arid tropics (SAT). This was agreed and AFNS recruited Hugh Doggett, one of the world's leading sorghum breeders who had spent more than 25 years working in the SAT of Africa. From the outset AFNS pursued a total systems principle in defining its programme and projects. Consequently, over subsequent years AFNS devoted significant resources to both agricultural production and post-production systems.

Production Systems

AFNS encouraged and supported production systems research in sorghum, various millets, grain legumes and oil seeds in north, west, east, southern and central Africa, and in the Middle East and Asia. Its support served to strengthen national sorghum and millet improvement programmes throughout francophone West Africa, and provided the first foreign aid grant to the newly independent Zimbabwe in 1980–1. The station at Matopos in southern Zimbabwe later became a regional sorghum and millet improvement centre.

IDRC was a founding member of the Consultative Group on International Agricultural Research (CGIAR) and AFNS acted on behalf of the CGIAR in creating the International Crops Research Institute for the Semi-Arid Tropics (ICRISAT). Since 1971 ICRISAT has maintained headquarters near to Hyderabad, Andhra Pradesh, and a diversifying programme throughout the world, mainly devoted to the improvements of sorghum, millets and groundnuts.

AFNS provided a team of four that recommended an integrated programme for Africa, which would have functioned from four strengthened national centres in Senegal, Nigeria, Uganda and Tanzania. Regrettably, the AFNS team's advice was not accepted and, eventually a new institute was constructed by ICRISAT in Niger.

AFNS' production research for crop improvement in the SAT sought always to strengthen indigenous capability, by linking universities and nations' agricultural programmes with ICRISAT, in addition to creating cooperative networks among several national SAT programmes. To this end, AFNS encouraged and supported a grain legumes cooperative programme among poorer nations located around the Mediterranean littoral. As a result, in 1975, CGIAR invited AFNS to act as executing agent in designing and creating an International Centre for Agricultural Research in Dry Areas (ICARDA).

It was originally intended that ICARDA undertake and coordinate research from three stations: (a) in the Bekaa Valley of Lebanon, (b) in the highlands of Iran, (c) close to the Euphrates in Syria. In consequence of various civil disturbances and changes in government, it was necessary to locate the headquarters and main station at Aleppo in Syria, with a highlands station in Turkey. Later, research resumed in the Bekaa valley. ICARDA continues to provide research support to cereal and legume producers throughout the drylands of West Asia, the Middle East and North Africa. Sorghums and millets were also developed in the Middle East as forage crops.

Sorghums originated in Ethiopia. To expand the range of germ-plasm available to sorghum breeders, AFNS financed sorghum germplasm collection missions to Ethiopia. Similarly, in cooperation with ICARDA, AFNS supported collection of chickpea in several West Asian territories from which chickpea is believed to have originated. By these means, AFNS enabled plant breeders to diversify the agronomic, functional and nutritional properties of these SAT crops.

It was AFNS policy to provide foreign 'expert' assistance only when essential resources did not exist within the SAT nations. A particular difficulty with certain genotypes of sorghum is the anti-nutrient polyphenols (tannins) biosynthesised in the pericarp and/or testa, the outer seed coats, of the grain. Many reported related studies assumed the sorghum polyphenols were hydrolysable tannins, similar to the substances extracted from oak galls (the ancient Celtic and Breton names for the oak were, respectively 'tannum' and 'tann') and other biological sources used since ancient Babylon in the tanning of animal skins to produce leather. AFNS identified one of the world's most experienced polyphenol chemists, Professor Edward Haslam of Sheffield University, provided him with a post-doctoral fellow, Dr Raj K. Gupta, who in due course demonstrated that sorghum polyphenols were condensed tannins related to many biosynthesised in the fruits and seeds of woody plants. This revelation made possible a more comprehensive understanding of sorghum tannins, their concentrations, functions and means of reducing their adverse anti-nutrient effects in breeding programmes.

Post-production Systems

It is essential in devising genotypes and phenotypes that generate higher yields (weight of edible material per ha/per yr) that the crops harvested be acceptable and usable by consumers for whom they are intended. From an extensive study in francophone West Africa, AFNS learned that many rural women devoted up to five hours every day to laborious pounding of sorghum, millets and dry legume grains into flour or meal. AFNS undertook an examination of grains structures to discover how the outer seed coats (husk and bran) could be mechanically separated from the endosperm, which when pulverised provides usable flour. Eventually, in cooperation with a farmers' cooperative in Maidugri, northern Nigeria, and Canadian agricultural engineers in Saskatchewan, a simple and effective milling machine was developed and installed in African rural mills.

The mill used rotating carborundum disks to remove the bran, air currents carried the bran out of the mill to be used as animal feed, the dehulled endosperm was collected and reduced to flour of various granularities in a hammer mill. A large group of young Nigerian women determined the properties of sorghum, millet and legume flour most acceptable to various consumer groups, the results indicating the quality of milled products most likely to be purchased. It was a firm AFNS principle that all post-production research be preceded by a consumer survey, to determine what properties were critical to acceptance and utility.

The sorghum milling activity became known to other African communities, with the result that AFNS supported small-scale rural mills in several nations. Carborundum tends to wear and to fracture when rotated above 1000 rpm, therefore, a later development replaced carborundum with lighter, thinner thermoplastic disks that could be rotated at up to 6000 rpm, thus increasing the durability and efficiency of the mills. For some 20 years, mills so constructed have been manufactured in several African countries. A study by a rural sociologist in Botswana discovered that their relief from many hours of pounding had enabled rural women to devote the time liberated to child care, the tending of kitchen gardens and the raising of poultry.

During the early 1970s, several donor agencies tried to persuade poor African farmers to construct small grain silos. The inherent weakness of all the proposals was that they relied entirely on materials to be imported from the donor country. AFNS encouraged a study of existing grain silos, constructed of locally available materials. Applying the recognised principles of harvesting and threshing before sunrise, when diurnal temperatures were lowest, drying by use of sun and prevailing winds, insulation with mud and protection of the silos from direct sunlight, grain sufficient for the annual needs of a typical rural West African farm family could be protectively stored in silos that cost only the local labour to construct them.

Sustaining R & D in SAT

It can be stated with some confidence that the AFNS programme in the SAT, started in the early 1970s, has been sustainable. ICRISAT and ICARDA continue to provide research and training support to many SAT nations, small grain mills continue to be constructed and to function in several African nations. Where rural programmes have collapsed or been debilitated, as in Zimbabwe, has largely resulted from government mismanagement and/or from civil or inter-communal strife.

The Integrated Systems Principle

The AFNS concept of integrated systems, to link production effectively with post-production, is now accepted among many organisations, not least among some of the world's largest transnational corporations who have created a Sustainable Agriculture Initiative, in which production is as much concerned with useful and nutritional properties as with high yield. Primary processing of perishable crops in rural areas, close to sites of production, serves to reduce post-production losses, a matter of considerable importance as cities become more congested and the time to transport harvested crops from rural areas to urban factories inexorably increases.

Social Forestry and Agroforestry

From past experience, several IDRC staff were persuaded that progress in agricultural development was more stable and sustained when technologists and artisans from similar ecologies were enabled actively to cooperate with one another than when left to struggle in isolation. One of the earliest illustrations of the concept was demonstrated in a collaborative social forestry network in which foresters from close to a dozen low rainfall nations in the Middle East, North, East and West Africa participated.

Gilles Lessard, the Associate Director of Forestry, whose academic qualifications were in natural resource management, convened an informal meeting of the first participants in Dakar, Senegal early in 1972. Each forester gave a brief account of his nation's perceived priorities and opportunities, the resources needed and available to pursue them. In all nations, soil conditions had been debased by excessive removal of forest cover. Trees variously provided fuel, fodder for animals, fruits and nuts, materials of construction, commercially valuable commodities such as gum arabic, an exudate of *Acacia senegal*, and shea butter, used in foods and cosmetics after extraction and refinement from seeds of *Butryospermum parkii*. Several of the participants noted that trees provided soil stability and soil nutrients: deep-rooted trees pump nutrients from sub-soils that are concentrated in leaves which, when they fall, enrich surface soils. In some nations trees acted as sand-dune stabilisers and protected crops from strong winds when planted as peripheral barriers. Some years later, an Egyptian forester, Hosny el Lakany, employed *Casuarina* species as windbreaks to protect crops in a successful project, financed by IDRC, that reclaimed many hundreds of hectares of desert land for crop cultivation and livestock husbandry. He had learnt their use in desert reclamation from Coptic monks who had long experience in dryland agriculture.

To facilitate frequent exchange of information and materials, IDRC employed two forestry advisors, Abdul Seif el Din, a Sudanese fluent in English and Arabic, and Karim Oka, an Algerian fluent in English, French, Arabic and Berber. The two foresters were given office space in Nairobi with funds to travel and convene meetings of the collaborators at least once each year. They provided a multilingual newsletter and when requested. IDRC provided monies to each of the collaborators to support priority projects. The social forestry programme eventually expanded to include some 20 projects supported by IDRC, most in arid and semi-arid ecologies of Africa and the Middle East.

In several of the francophone nations all forests and tree lots were owned by the government departments of Eaux et Forets. Trees and water owned by governments were frequently ill-protected and regarded as free to be exploited by all and sundry. Consequently, villagers harvested whatever trees and arboreal products they needed, while significant quantities were scavenged by goats and other migrant animals. It was suggested, and in two instances accepted, that all woodlots be declared the property of the nearest village community. As soon as the villagers owned the woodlots, they took pains to prevent depredation by animals and engaged in planting, protecting and conserving their woodlots efficiently and economically. In several locations, IDRC supported tree nurseries to generate species of particular utility to different ecologies and communities.

Other projects supported training of young rural people and establishment of small industries in timber processing including, in Tunisia, the production of paper from woody plants indigenous to the local ecology; in Mali the hybridisation of mangoes, and the establishment of small sawmills.

Agroforestry

During the 1970s, interest arose in integrating forestry with agriculture, alternatively named agrosylviculture or agroforestry. Soon after IDRC was created, there was pressure from some politically influential persons in the Canadian forest industry for the Centre to finance an international institute of forestry research in Canada. Eventually it was agreed that IDRC would explore the feasibility and utility of supporting an international programme to integrate forestry with crop and livestock production. Several international development agencies expressed interest and, in the later 1970s, after a somewhat difficult period, largely the result of inept politically influenced interference, a Centre for agroforestry research was established with headquarters in Nairobi and a small Board of Trustees drawn from several nations.

At first the organisation, named the International Council for Research in Agroforestry (ICRAF), followed the pattern of the Social Forestry network by providing advice mainly to African nations on various opportunities by which to cultivate and utilise arboreal species to protect food crops, to provide fodder for animals, to act as wind and insect breaks, and to provide soil nutrients in the form of leaf mulch. A search committee, led by a former Director General of the International Rice Research Institute, identified an exceptionally competent Director, Bjorn Lundgren of Sweden, and recommended that ICRAF begin its own research programme on land provided by the Kenya government. Given the qualifications of Gilles Lessard, it is interesting that the Director described ICRAF's programme as research and development to utilise land and related natural resources most efficiently.

During the 1980s other donor agencies provided support to ICRAF and the organisation was accepted as an associate member of the CGIAR. According to CGIAR records, ICRAF became a full member of the CGIAR in 1991. ICRAF is now a truly international research centre with programmes and projects in Africa, Asia and Latin America. Nonetheless, ICRAF metamorphosed from the IDRC social forestry programme and it was IDRC that gave it birth and nurtured it during its early, difficult years.

Local Management of Water

In the précis of his book *Water; Local-level Management,* David E. Brooks gives blunt advice: "There is one iron rule for managing groundwater and aquifer supplies: assume the worst". And then, from a man who has been in the forefront of this field heading an NGO (Friends of the Earth) and worked as a consultant before joining IDRC, Brooks adds more positively: "Policy and research should shift their focus from enlarging supplies of water to managing demand". On the question of engaging local communities, he says: "Policy-making should always start by accepting social customs and cultural norms as given, but not sacrosanct". Finally a warning: "Devolution of the power to manage water (not just read meters and fix leaks) will not come easily. The forces to maintain a top-down approach to water are well entrenched and serve many power elites".

His way of thinking has informed a wide range of water projects supported by IDRC, that have taken place at all levels from international to community. Internationally, the Centre has worked for a dozen years in the Middle East and North Africa region to promote water demand management (WDM) through gatherings of ministers and officials to exchange knowledge and experience among their nine regional countries. Missions between countries have shown, for example, the Syrian water officials how Water Users Associations (local bodies) in Tunisia distribute drinking water and levy charges, and how Egypt deals with wastewater.

These gatherings have evolved into a Water Demand Initiative, a five year-programme (2004–2009) with the awkward acronym WaDImena, to reach out further to civil society with research, and to apply WDM strategies in particular rural areas, concentrating on women and the rural poor. WaDImena has a website (www.idrc.ca/WaDImena) with a trilingual glossary of WDM terms and a library that highlights lessons learned.

But respect for traditional knowledge is a major factor in tackling water demand problems. Farmers in the stony highlands of Yemen depended for years on intricate terracing to save water, and conserve fertile soils from erosion. Researchers studying why these terraces were now in disrepair learnt the obvious answer: maintaining them was brutally hard labour, and many men had left for the cities. Also, there was no clear agreement on cost-sharing between landlords and tenants, and no credit available to farmers to invest in water management. Ways were found to rebuild at reasonable cost and labour, and food production became profitable and attractive again. A comment from David Brooks: "Reviving traditional water management approaches can require both technical and policy ingenuity, but the rewards can be significant".

China provides the most striking example of how a determined (and sometimes desperate) community can take on management of its water resources, and reverse the top-down process of regulation. Centralised state control enforced by a cumbersome bureaucracy has long been the pattern in China, and one recent result has been that the economic boom in its eastern provinces have brought little benefit to parts of the interior. In particular, the mountainous Guizhou province in the southwest has remained one of the poorest; and Changshun county, set on sloping land that is mainly porous limestone is chronically short of ground and subsurface water. Shortage of water has meant low rice yields, little diversification of crops and degraded forests. It has added to the burden of women who walk miles to collect water for households.

A first attempt at water management was made in the early 1990s when facilities were rebuilt and maintained by the state. But there was little accountability of this poorly managed government project, and no local control. Politely put, the project had limited effectiveness. At that point, researchers at the Guizhou Academy of Agricultural Sciences (GAAS), funded by IDRC, decided to try a community-based approach, involving participatory decision-making about natural resource management. In 1995 its team began learning about and analysing traditional practices in a number of villages, and measuring the damage already done to the natural resource base there. (It was the same diagnostic approach described in the case-study on Farming Systems Research).

The GAAS team guided the academics and local residents through a collaborative process. There was debate on practical and technical matters – irrigation, reservoirs, pipes, new water sources – and strong attention to the social aspects of development. But the overall emphasis was on the *process* and on involving people in decisions about their own development. It was painstaking and lengthy, but the process worked. The people selected the technical solutions, and built the water systems; they took 'ownership' by working out regulations and agreed payments for maintenance and by electing (and paying) a manager to run daily operations.

Results reported have been remarkable. Farmers diversified crops and enjoyed increased yields. They grew fruit trees and berry bushes on the marginal land. Women probably gained the most. Relieved of hours of water gathering (it was now piped, a tap to every 50 households), they took charge of a peach orchard, and marketed the fruit. They overruled the researchers, who had suggested chestnut trees; and they impressed the deputy governor of Guizhou province, who had expected the scheme to fail, and who later admitted his officials had not thought of the kinds of regulations and management systems the villages established. The idea that local initiatives can solve local problems is gaining ground in China. David Brooks should have the last words: "Decision-makers too often dismiss small groups and small solutions. They make a big mistake".

Essential Health in Tanzania

There are some extraordinary, in fact appalling, statistics cited about Tanzania. Its national income in 2001 of $9 billion, divided among 35 million citizens, was roughly half what Americans spent that year on wallpaper. Its spending on health that year was US$11.37 a head, compared with $2,809 in Canada and $4,819 a head in the United States. A 'poverty head count' in 2003 showed 36 per cent of Tanzania's population living on less than US$1 a day. And there were an estimated 7,431 Tanzanians to every (ill equipped) health facility.

After independence in 1961 the government under President Julius Nyerere and his people made heroic efforts to build a health care system throughout the country. Under a unique social contract the rural people overcame their reluctance to move into modern villages, and the government provided pumped water, a clinic and a school. Graduates spread into rural areas, medical training centres were built. Health care was a public right.

By the mid-1980s the system was foundering. Its centrally planned management was unable to sustain the structure and to support efficiently the village dispensaries. And the people were coming under renewed assault from infectious diseases – malaria, tuberculosis and now AIDS. Payment obligations on international debt, coupled with sharply falling prices in export crops, cut deeply into funds for training health staff. Attempts to keep the system going by charging user fees for low-quality care only drove people angrily away from the whole system.

Even when the UN Global Fund and charities such as the Bill and Melinda Gates Foundation pledged substantial funds for health care in Africa, the health systems of Tanzania and other countries were weakened and had little ground-level capacity. The *World Development Report 1993: Investing in Health* [World Bank, 1993] argued that new investments should be based on evidence about the local "burden of disease" and be targeted proportionally. Increase health funding, certainly; but make sure it is allocated in a cost-effective way. The logic was good, but how to put it into practice?

Canada through the International Development Research Centre contacted several African countries to suggest a partnership to research this problem. Tanzania, intent on health care reform and planning to devolve the responsibility of budget management and health care delivery to local authorities, was the first to respond. The Tanzania Essential Health Interventions Project (TEHIP) has evolved as a true partnership between IDRC and the Ministry of Health, with Don de Savigny and Conrad Mbuya managing and coordinating research, and Graham Reid and Harun Kasale managing and coordinating the actual project.

TEHIP was launched in October 1993 with careful, broad-based consultations to design the project. There followed the stage of discovering how local authorities could get an accurate picture of the existing 'burden of disease' – which diseases imposed the greatest cost of lives (and years of life) lost in their area. It chose two rural districts, Morogoro and Rufiji, west and southwest of Dar es Salaam, with a total population of 741,000.

Most health data in Tanzania is admittedly inaccurate, being collected from clinics – while most people die in their homes. TEHIP therefore, sent researchers out on bicycles on a door-to-door survey, asking representative households whether anyone had died or been laid low with sickness recently, and with what symptoms. (Again the diagnostic approach: first, go and get the facts and views from the people affected). The findings were striking, showing some diseases were badly neglected. Malaria accounted for 30 per cent of the years of life lost in Morogoro, but had only 5 per cent of recent health budgets. Conversely, tuberculosis accounted for less than 4 per cent of years of life lost, but received 22 per cent of the budget.

IDRC funding offered a modest $2 a head top-up to the districts' budgets, but with re-allocation of existing funds only a part was at first needed in the transition to a more cost-effective system. Simple rules were followed: the cheapest treatments were offered first; drugs were ordered according to what was needed. (Under the old centralised system, all dispensaries received the same package of pills). Most important was the campaign to persuade people to use bed-nets impregnated with insecticide (these cost $3, locally made).

TEHIP remained in the background as the District Health Management Teams (DHMTs) planned their work and spent their money. The teams could work more efficiently unburdened with volumes of charts and paper, but armed with what Don de Savigny calls a "tool kit": a district profile of the 'burden of disease' with key indicators, graphs to cover trends in spending areas, health maps, community reactions and ideas, cost-tracking – all providing as simply as possible on computer diskettes.

The results were swift and dramatic. Adult mortality has dropped by nearly 20 per cent. Infant mortality fell by 28 per cent in one year in Rufiji, and the proportion of children dying before their fifth birthdays by 14 per cent, and in Morogoro annual under-5 mortality fell from 35 to 20 per 1000 within a few years. The teams achieved this by adopting another 'tool' that was devised by WHO: the Integrated Management of Childhood Illnesses (IMCI) which meant (as de Savigny says) "addressing the whole child by identifying and treating a range of possible common illnesses rather than simply focusing on one disease at a time".

It can soon be more than a local achievement. DHMTs have been trained in half the country in the use of the burden of disease profile and the district health accounts tools. The United Nations has funded the 'roll-out' of the entire tool kit to 11 other Tanzanian districts, and CIDA has chipped in $7 million CAD to build on the results of TEHIP. Further afield, the Danish aid agency DANIDA is testing it in Ghana, and South Africa is showing a general interest in the TEHIP approach.

Bibliography

Ausubel, F.M.: *Current Protocols in Molecular Biology*, New York, John Wiley and Sons, 1989.

Brandwein, P.F.: *Sourcebook for the Biological Sciences*, San Diego, Harcourt Brace JOvanovich, 1986.

Brooks, Rodney A.: *Flesh and Machines: How Robots Will Change Us*, New York, Pantheon Books, 2002.

Chrispeels, Maarten : *Plants, Genes and Crop Biotechnology*, Sudbury MA, Jones and Barlett Publishers, 2003.

Cox, Earl D. : *Beyond Humanity: Cyber Evolution and Future Minds*, Roackland, Charles River Media, 1996.

David Sadava: *Plants, Genes and Crop Biotechnology*, Sudbury MA, Jones and Barlett Publishers, 2003.

Dixon, Dougal: *After Man-A Zoology of the Future*, New York, St. Martin's Press, 1981.

Dodds, John H.: *Plant Genetic Engineering*, New York, Cambridge University Press, 1985.

Drexler, K. Eric: *Engines of Creation: The Coming Era of Nanotechnology*, New York, Anchor Books, 1986.

Fransman M, Junne G, Roobeek A: *The Biotechnology Revolution?*, Oxford, Blackwell, 1995.

Friedberg, E.C.: *DNA Repair*, New York, WH Freeman and Company, 1985.

Fumento, Michael: *Bioevolution: How Biotechnology is Changing Our World*, San Francisco, Encounter Books, 2003.

Gaskell G: *Biotechnology-the Making of a Global Controversy*, Cambridge, Cambridge University Press, 2002.

George S. Paul: *Beyond Humanity: Cyber Evolution and Future Minds*, Roackland, Charles River Media, 1996.

Goodman, D.C.: *From Farming to Biotechnology: A Theory of Agro-industrial Development*, Oxford, Blackwell, 1987.

Goodsell, David S.: *Bionanotechnology: Lessons From Nature*, Hoboken, Wiley-Liss, 2004.

Kurzweil, Ray: *The Age of Spiritual Machines*, New York, Penguin Books, 1999.

Moravec, Hans: *Mind Children: The Future of Robot and Human Intelligence*, Cambridge, Harvard University Press, 1988.

Morholt, E. : *Sourcebook for the Biological Sciences*, San Diego, Harcourt Brace JOvanovich, 1986.

Murray, David: *Seeds of Concern: The Genetic Manipulation of Plants*, Sydney, University of New South Wales, 2003.

Old, R.W. : *Principles of Gene Manipulation*, London, Blackwell Scientific Publications, 1989.

Postman, Neil: *Technopoly: The Surrender of Culture to Technology*, New York, Vintage Books, 1992.

Primrose, S.B.: *Principles of Gene Manipulation*, London, Blackwell Scientific Publications, 1989.

Retzer, W.J.: *Biotechnology Workbook*, Englewood Cliffs, Prentice Hall, 1991.

Rifkin, Jeremy: *The Biotech Century*, New York, Penguin Putnam, 1998

Shetty, Kalidas: *Food Biotechnology*, New York, Dekker/CRC Press, 2005.

Slater, R.J.: *Experiments in Molecular Biology*, Clifton, Humana Press, 1986.

Smith, John E.: *Biotechnology*, Cambridge, Cambridge University Press, 2004.

Switzer, R.L.: *Experimental Biochemistry*, New York, W.H. Freeman and Company, 1977.

Towle, Albert: *Modern Biology*, Austin, Holt, Rinehart and Winston, 1988.

Walden, Richard: *Genetic Transformation in Plants*, England, Open University Press, 1988.

Winston, Mark L.: *Travels in the Genetically Modified Zone*, Cambridge, Harvard University, 2004.

Index

P

R

S

T

U

□□□